GEM AND ORNAMENTAL MATERIALS OF ORGANIC ORIGIN

To Louise and Kathrine, with my love

GEM AND ORNAMENTAL MATERIALS OF ORGANIC ORIGIN

Maggie Campbell Pedersen
FGA, ABIPP

AMSTERDAM BOSTON HEIDELBERG LONDON NEW YORK OXFORD
PARIS SAN DIEGO SAN FRANCISCO SINGAPORE SYDNEY TOKYO

Elsevier Butterworth-Heinemann
Linacre House, Jordan Hill, Oxford OX2 8DP
200 Wheeler Road, Burlington, MA 01803

First published 2004

All photographs by the author, unless otherwise stated

British Library Cataloguing in Publication Data
A catalogue record for this book is available from the British Library

ISBN 0 7506 58525

For information on all Butterworth-Heinemann publications
visit our website at www.bh.com

Typeset by Genesis Typesetting Ltd, Rochester, Kent, UK
Printed and bound in Italy

Contents

Introduction

Gem and ornamental materials of organic origin – the so-called 'organics' – belong in a class of their own. The first jewellery was made of organic materials, long before anyone thought of gem minerals. Some seeds or nuts on a string – and possibly a few feathers – were very decorative and were easily accessible. They needed no preparation so could be used by the most primitive of peoples. It is likely that this form of adornment was used before humans had worked out how to make clothes.

Nothing went to waste in the lifestyle of our ancestors, so the bones, claws, teeth and even whiskers of anything that was hunted were not discarded, but were put to use. The largest items such as tusks or large bones could be used as building materials, while the smallest items like claws and teeth were worn as adornments. The hunter that wore the most teeth was obviously the one that caught the most animals, and so the 'jewellery' became a symbol of status.

Through the ages organics have been used for trade and barter, and as hard currency. They have been used as offerings to the gods, and as gifts laid in a burial mound to be taken to the life hereafter.

Organics have a rich background – to my mind far more fascinating than that of minerals. They have been involved in some dark areas of our history, such as the slave trade, when ivory and slaves were often traded together. The illegal collection of some of the organics – for example, Baltic amber – has, in the past, carried the death penalty. Even today, ivory poachers risk being shot.

A favourite use of organic materials has always been in the form of ceremonial attire. Feathers took a prominent position in this and were made into capes, necklaces and head-dresses. Examples can be found worldwide. Animal teeth, claws, scales, bones and shells were often combined with the feathers. Ceremonial and hunting masks have had widespread use, from the Inuit masks of wood with ivory and sealion whisker adornment, to the imaginative tortoiseshell masks of the peoples of the Torres Straits on the other side of the world.

Many of the organics have been perceived as having talismanic and amuletic powers, and quite often the same powers were attributed to different organics. Several of them were thought to neutralise poison, and to ward off certain illnesses. Taken in powdered form, many of the organics were prescribed for the same ailments, such as chest complaints or rheumatism.

Among the various reasons for grouping the organics together is that they complement each other and are aesthetically pleasing when used in combination. For example, mother-of-pearl and tortoiseshell have been combined as inlay, tortoiseshell and ivory as veneer, and coral and pearl as jewellery.

Gem materials are supposed to follow the three basic principles: that they are beautiful, rare and durable. Organics are usually beautiful, often rare, but they are not necessarily durable. They are much softer than minerals, and are easily damaged. Some organics fade in sunlight and others darken, yet even with these drawbacks organics have always been held in high esteem. They tend to be very tactile, and quite often their lustre is improved by constant handling.

Another unique feature of organics is that some of them are still being produced. Although ambers and jet are, like minerals, millions of years old, other organics such as coral and pearls are growing today and may be harvested tomorrow. There are sheep and cattle growing horns, and stags growing antlers.

A peculiarity common to all the organic gem materials is that they are very hard to test by normal, gemmological means, and are best identified by sight. This book attempts to address that problem, and to give guidelines for identification through the numerous photographs.

I have tried to avoid the popular terminology used for some of the materials. For example, Baltic ambers have various descriptions in different countries, such as 'bony' or 'turbid', and a certain colour of coral is sometimes called 'angel skin'. I have felt it best to rely on descriptions instead of the popular names which can be confusing. Similarly, the words 'imitation' and 'simulant' are used instead of the popular expression 'faux'. The exception is the word 'organics', which I use freely, instead of the more correct, but rather long, 'materials of organic origin'.

It has not been my intention to go into very great depth about each subject covered in the book, but it is hoped that it will act as an introduction to the whole concept of organic gem and ornamental materials, and how they are identified. However, I believe that it is not only interesting, but also important, to know where the materials come from, so I have attempted to give some of the background of each. This has meant that I have touched on subjects about which I have limited knowledge. I have endeavoured to ensure that the facts given in the book are correct, and to that end I have consulted numerous experts. Even so, there may be some errors. Any such errors are mine, and for them I apologise.

There is frequent mention in the book of CITES, which is the Convention on International Trade in Endangered Species of Wild Fauna and Flora. It is an international agreement between governments which aims to ensure that the survival of wild animals and plants is not endangered by trade in any of their parts. At the time of writing, there are 160 member states which undertake to respect the bans laid down. Great Britain is a signatory to the agreement, with the effect that we cannot sell any flora or fauna that may be listed as endangered in our country, nor can we import any listed items from other countries, whether those countries are signatories to the agreement or not.

CITES meets every two years, and revises the three lists with which they operate:

> Appendix I includes species threatened with extinction. Trade in species on this list is only permitted in very rare circumstances.
> Appendix II covers species that may be threatened with extinction if their trade is not strictly controlled.
> Appendix III covers species that are protected in at least one country, where assistance is needed to control the trade.

Many of the organic materials mentioned in this book are covered by trade bans and conservation treaties. Some countries are not signatories to CITES yet have their own trade bans. It is therefore not possible to give completely up-to-date lists, as these can quickly change. It is advisable that anyone seeking to purchase any of the organic materials first checks their status, and if necessary obtains the appropriate legal documents.

Maggie Campbell Pedersen
April 2003

Acknowledgements

If I were to name everybody that has helped me with the preparation of this book, the Acknowledgements would be as long as the book itself. I have met with unlimited help from friends and acquaintances, and even from people with whom I had never spoken before. I have received a lot of encouragement, and a tremendous willingness to share knowledge. Because of the broad scope of this book, I have had to appeal for help and advice from very many people. The response has been amazing.

People have given generously of their time to aid me in my quest for a deeper understanding of the subjects. From jet workers in Whitby and horn workers in the Lake District, to people at the Environmental Investigation Agency and those working tirelessly on hawksbill conservation at the Barbados Sea Turtle Project – all have been willing to help.

Many individuals have also given me assistance. Among them are dr.phil. Jørgen Jensen, Mogens Bencard, Stig Andersen, Dr David Grimaldi, Søren Fehrn, the late Patricia Lapworth and countless others. Constant help has come from Doug Garrod, Lorne Stather, and everyone at the Gemmological Association of Great Britain (Gem-A), who have researched, discussed, and generally always been there for me. Michael O'Donoghue has given me great support and good advice over many years.

The following people have proof read various parts of the manuscript. Some have read a few pages related to their own speciality, while others have read through and corrected several chapters. They include: Ian Mercer (who has also helped me with many identifications over the years, and whose patience is unlimited), Mary Burland and Stephen Kennedy at the Gemmological Association of Great Britain; Gary Jones, Richard Sabin, Jill Darrell, Dr Brian Rosen and Professor John Taylor at the Natural History Museum in London; Dr Paul Jepson at the Institute of Zoology in London; Dr Julia Horrocks of the University of the West Indies; Christine Woodward; Sylvia Katz; and

Adèle Scheverien, archivist to the Horners Company. My sister, Alison Campbell-Jensen, has read most of the manuscript – the book would have made little sense without her help with syntax. Promises Specialist Colour Processing Laboratory in West London have processed the film – with the exception of the photographs lent by museums – for all the photographs in the book.

My family, especially my two daughters Louise and Kathrine, have also been unstinting in their reassurance over the long period of time it has taken to complete the task. Finally, mention should also be made of my late parents: my father, who first introduced me to gems and helped me look for amber on a Danish beach when I was seven years old, and my mother, who had the unquestioning faith in my abilities that only a mother can have.

I am indebted to all of the above. My sincere thanks to them, and to the many more who are not mentioned by name.

Photo credits

All the photographs in the book were taken by the author, with the exception of the following, which were kindly lent by museums. The copyright is retained by the museums.

Fig. 1.37 The National Museum of Denmark.
Fig. 1.39 The Royal Collections, Rosenborg Castle, Copenhagen, Denmark.
Fig. 3.19 The V&A Picture Library, London. (photography: Mike Kitcatt).
Fig. 3.7 The Royal Collections, Rosenborg Castle, Copenhagen, Denmark, (photography: Bent Næsby).
Fig. 6.3 The V&A Picture Library, London. (photography: Ian Thomas).
Fig. 7.12 The National Museum of Denmark.
Fig. 8.16 The British Museum.
Fig. 9.23 The V&A Picture Library, London.
Fig. 11.25 The Royal Collections, Rosenborg Castle, Copenhagen, Denmark, (photography Kit Weiss).
Fig.12.12 The Royal Collections, Rosenborg Castle, Copenhagen, Denmark, (photography Kit Weiss).

The items in the following photographs were kindly lent to the author for photography, by the Gemmological Association of Great Britain, Gem-A.

Figs 3.13, 3.15, 3.17, 6.2, 9.9, 10.1, 12.11.

All other items are in private collections, unless otherwise stated.

Notes on the tests used in this book

Unlike the inorganic gems, most gem and ornamental materials of organic origin are not routinely tested using the normal gemmological tests. For example, a common gemmological test involves taking a reading of the refractive index of the material, by putting a drop of contact liquid onto a flat surface of the material, and placing it in a small instrument called a refractometer. Many of the items dealt with in the book are too large for this test, and those that are small enough would probably not have surfaces sufficiently flat for a reading to be taken. Also, the liquid used can cause damage to the surface of most organics. Similarly, specific gravity and hardness tests are usually inappropriate for these materials. Organics, by their very nature, defy the strict boundaries of normal gemmological tests.

'Observation' is the key to identifying organics. Examination in a good light – a cold light source is useful for this – possibly with the help of a magnifying lens such as a 10×, will usually give a strong indication of their nature. Sometimes it helps to look at them under stronger magnification, with a microscope. Occasionally feel and even smell can help with their identification.

In the event that there is still doubt about an identification, a few simple tests can be carried out, which are described at the end of each chapter. It should be noted, however, that some of the tests are destructive to the specimen and involve such actions as taking a tiny scraping of the material.

There are three tests that, if carried out incorrectly, can also involve some danger to the person executing them.

1. The **hot-point test** involves pressing the heated point of a needle or pin against a hidden surface of the material – for example, inside the drill hole of a bead – to gauge the result. *If the material is made of celluloid, this action can cause it to combust.* The test can also damage some materials that burn easily and thereby leave a large hole in the specimen. The test cannot give definitive results, but,

when several specimens are tested at the same time and the results are compared, it can give some indications.

2. The **burning test** involves taking a tiny scraping of the material and is a safer version of the above test, but again care must be taken as even a scraping of celluloid *can combust*. It has the disadvantage that it is hard to judge the speed at which a material such as a natural resin will burn (which can give an indication of the nature of the material), whereas it is easy to see the speed with which a hot needle melts the material.
3. **Ultraviolet (UV) light** is not so easily accessible, and the tests are usually carried out in controlled circumstances, so there is not a warning at the end of each chapter as there is for the burning tests. However, it must be noted that UV light can be dangerous to health if not treated with due respect. The eyes and skin should not be subjected to direct rays.

This book refers to long wave UV light for, although gemmological tests include exposure to both long and short waves, most organics react little – if at all – to short wave. An advantage is that long wave UV light is marginally safer to use.

UV light is not destructive to the specimen and is very useful as it can often help to determine the difference between a natural material and a plastic simulant. It is not infallible though, as some materials – especially amber – lose some of their fluorescence, or fluoresce with a different colour, as their surfaces oxidise, either from age or from heat treatments.

Inevitably there are some items which it is not possible to identify without expert help. An example of this is pearls, where it can be impossible to judge whether they are natural or cultured. Gem testing laboratories are equipped to deal with these queries, and are able to carry out the more sophisticated tests that may be necessary, such as X-ray or infrared spectroscopy.

1 Amber and copal

Amber and copal are tree resins, which are materials produced by some trees as a form of protection. Resin usually runs down the trunk under the bark and seeps out if the bark is damaged or broken. Resin should not be confused with sap, which is a nutrient-bearing substance contained in all plants.

There are many resins that can resemble amber and copal, but these are the only two that can be termed 'gem materials'.

Put very simply, copal is young version of amber. There is no definite age at which copal turns into amber, as the process is continuous from the moment the resin appears on the tree and begins to solidify. In physical terms, when the resin is sufficiently cross-linked and polymerised it becomes amber (see Chapter 13, 'Plastics'). In other words, the resin has dried out and hardened. This process takes thousands if not millions of years, and not all copal becomes amber as much of it disintegrates with time. Furthermore, as the process is such a long one it is not possible for us to follow it or to replicate it in a laboratory, so there is still much that is speculation. We know, however, that there are some instances of copal that have begun to look like, and take on, the properties of amber.

Ambers

Amber deposits are found worldwide and are reckoned to be from 300 to 15 million years old. The two best known deposits are those in the Baltic region and in the Dominican Republic.

There are few signs of the forests that once produced the original resin as they are long gone. They have been washed away with the changing land masses of the planet, and died off as the climate became colder in the north.

Amber is too old for carbon dating, so one method used for determining its age is by dating the geological layers in which it is found.

However, this method can be confused by the fact that many amber deposits are secondary, for example much of the Baltic amber was moved to its present location by melting ice between past Ice Ages. Burmese amber was long thought to be about 45 million years old, but recent research suggests that it probably originated elsewhere and not in its present location. It is therefore a secondary deposit, making it much older, at between 100 and 80 million years old. This conclusion was reached by studying the insect inclusions, which did not match the age of the rock strata in which it lies.

While many gem materials contain inclusions, amber is unique in that it can contain inclusions of plant and animal matter – flora and fauna. These have become stuck to the sticky surface of the resin as it oozed from the tree. They may have inadvertently landed on it, or they may have been blown onto it by the wind. In the case of stingless bees, it is thought that they gathered the resin for nest building. Other trapped creatures have included termites, ants, flying ants, butterflies, moths and damsel flies, or anything too small or weak to be able to extricate itself.

Plant inclusions are those that were light enough to be blown on the wind, or picked up on the forest floor at the base of the tree when the resin reached that far. There have been very rare findings of tiny frogs or scorpions. As time passed the insect or plant matter became encapsulated in the substance which, due to its chemical properties, dried out the object without shrinking it, sterilised it and finally hermetically sealed it for us to examine millions of years later. It is because of this preservation that we are able to learn much of what the world was like all those years ago, long before man emerged on this planet (Figs 1.1–1.5).

Baltic amber

Known as the Gold of the North, Baltic amber is the best known of all ambers. It has the longest history and is the most plentiful.

Around 50 million years ago Scandinavia and the Baltic formed one land mass covered in forests. The climate was subtropical. It is here that Baltic amber was formed between 45 and 30 million years ago. It possibly came from different trees, though the one which is popularly thought to have produced the resin has been named *Pinus succinifera*, as it bore fruit resembling pine cones and the resin contains succinic acid.

Deposits occur in the eastern Baltic from Poland to Estonia, especially around the Samland Peninsular. Here it is mined in large quantities.

Baltic amber is found as secondary deposits in much of Denmark and southern Sweden, in layers laid down 30 000 years ago. It is not mined but turns up when foundations are laid for bridges or buildings,

Figure 1.1 Insect in Baltic amber, showing typical white coating.

Figure 1.2 Fungus gnat in Baltic amber.

Figure 1.3 Flower head in Dominican copal.

Figure 1.4 Damsel fly in Dominican amber.

Figure 1.5 Poorly preserved bark lice in burmite.

or it is washed up on the beaches after storms that dislodge the material from the seabed. Occasionally it also washes up on the north Norfolk coast of England.

Structure and properties of Baltic amber

- Baltic amber is the most varied in colour and transparency, ranging from creamy opaque through transparent golds to almost black, and often occurring as a mixture of different shades and transparencies (Fig. 1.6).
- Amber consists basically of carbon, hydrogen and oxygen, with traces of other elements. Baltic amber contains a much higher percentage of succinic acid than any other amber and is therefore sometimes named 'succinite'. (To some purists, only 'succinite' is real amber. Other ambers, which contain no succinic acid, are then called 'retinites'.)
- Rough amber has a dull, matt appearance and unless broken shows little of the golden glow of the polished article.
- Amber breaks with a conchoidal fracture. Usually a single lump will contain a variety of colours and opacities (Fig. 1.7).
- The opaque appearance of some amber is caused by minuscule air bubbles trapped inside the material – the smaller the bubbles the

Figure 1.6 Pieces of Baltic amber rough and half-polished, showing variations in colour.

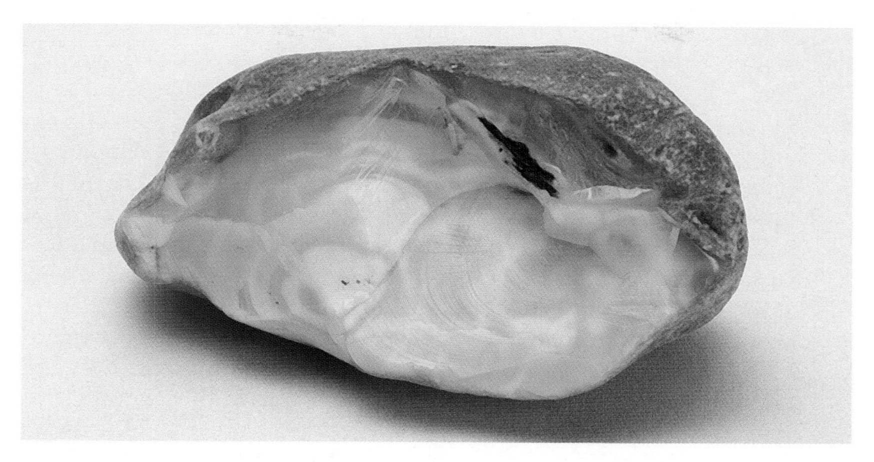

Figure 1.7 Piece of Baltic amber rough, showing different opacities and conchoidal fracture.

Figure 1.8 Baltic amber beads showing different opacities.

more opaque and paler the material. One square millimetre of opaque amber can contain anything from 600 to 900 000 bubbles (Fig. 1.8).

- Baltic amber is rich in inclusions and over 1000 varieties of flora and fauna have been discovered embedded in the resin. Unique to Baltic amber are hairs from oak flowers and tiny crystals of the mineral pyrite. Other inclusions are flowers, leaves or pieces of bark, and various insects, the most common being termites and ants. Most of these things are now extinct, but a few species are very closely related to insects or plants alive today (Fig. 1.2).
- An opaque, white matter is often present around an insect inclusion. This is thought to be a fungus formed by the partial decay of the insect, and is unique to Baltic amber (Fig. 1.1).
- Baltic amber fluoresces pale, chalky yellow to chalky pale blue under ultraviolet light. The opaque areas fluoresce more than the transparent areas.

Dominican Republic

The second most plentiful amber, that from the Dominican Republic, is younger, at between 25 and 15 million years old. Some of it is believed to be the resin from a shrub of the genus *Hymenaea*, but several different types occur, with varying ages and different physical properties.

This amber is also mined, occurring in veins of lignite in mountainous areas.

Structure and properties of Dominican amber

- Dominican amber looks very different from Baltic amber. It is almost always transparent and of yellow to reddish-brown hues, or occasionally a pale greenish-yellow. It contains very large quantities of black substances picked up as it formed (Fig. 1.9).
- It is very rich in inclusions that are magnificently preserved, and are usually clearly visible in the transparent resin (Fig. 1.4).
- It contains no succinic acid and is sometimes termed retinite. It can show strong green, blue or beige fluorescence under ultraviolet light. The green or blue can also sometimes be seen in daylight.

Mexican

Mexican amber comes from the Chiapas region. After being unobtainable for some years, it is again appearing on the market, though in small quantities only. It is between 30 and 20 million years old and, like Dominican amber, it originates from a tree of the genus *Hymenaea*. It is much sought after and popular for use as a material from which to make good-quality carvings.

Figure 1.9 Pieces of rough amber from the Dominican Republic.

STRUCTURE AND PROPERTIES OF MEXICAN AMBER

- Mexican amber is almost always transparent, occurring in colours from pale yellow to deep, rich, red-brown. It seldom contains flora or fauna inclusions, though it often contains lots of dark debris (Fig. 1.10).
- Some Mexican amber contains typical, almost parallel strata of dark debris (Fig. 1.11).
- By transmitted light, some Mexican amber appears to have areas of rich red colour, but it does not resemble the clear, burnt-orange to cherry red of burmite (Fig. 1.10).
- It can display strong green to blue fluorescence in daylight, which appears to come from the body of the amber rather than the surface. It has a chalky blue or chalky beige surface fluorescence under ultraviolet light (Figs 1.12 and 1.13).

Burmite

Burmite comes from the Hukawng Valley in Kachin State, in northern Myanmar (formerly Burma). Originally thought to be younger, it is

Figure 1.10 Polished Mexican amber pieces, showing variations in colour.

Figure 1.11 Polished Mexican amber showing parallel lines of debris (magnified).

Figure 1.12 Frog carving in Mexican amber.

Figure 1.13 Frog carving showing fluorescence in daylight.

now realised that it is from the Cretaceous period, and dated at between 100 and 80 million years old. This makes it by far the oldest of the popular ambers, as all the rest are from the Tertiary period (65 million years ago to the present day). Of all the ambers, burmite is thought to contain the greatest variety of insect inclusions, though many of these inclusions are not well preserved and little more than the outline of the insect remains (Fig. 1.5).

For many years it was not extracted due to the political situation in Myanmar, but a small amount is again being mined and is available on the market. Burmite is a very beautiful material, which has for centuries been prized by the Chinese for carving. It has frequently been copied, either by dyeing the surface of amber from another source, or by reconstituting amber and adding a red dye. It was also copied in red plastic, for example in Bakelite (Figs 13.5 and 13.6).

Structure and properties of burmite

- Burmite is harder than other ambers and can contain many insect inclusions.
- Burmite takes a very high polish.
- It can be very clear and uniform in colour. It can also vary from cream and yellow to dark brown and rich red, and it may have striking swirls of these colours (Figs 1.14 and 1.15).

Figure 1.14 Pieces of burmite, rough and half-polished, showing variations in colour.

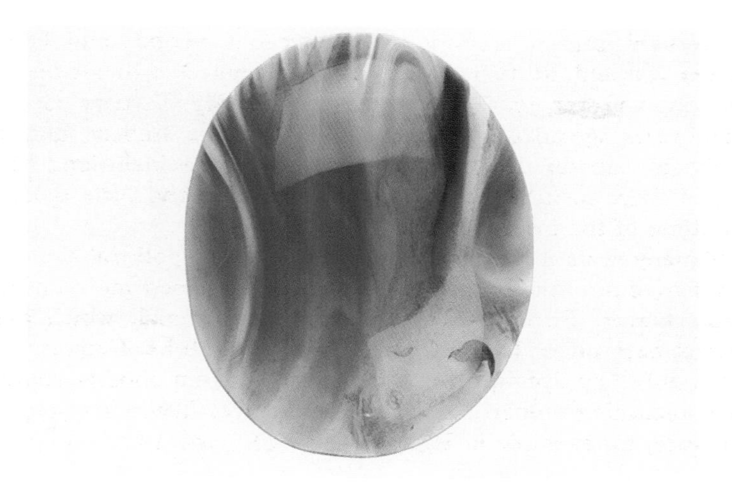

Figure 1.15 Burmite cabachon, showing typical swirls of colour (magnified).

Figure 1.16 The same burmite cabachon under ultraviolet light.

- When viewed under high magnification, these swirls appear as microscopic spots of colour (Fig. 1.17).
- Turning a piece of burmite in the light often gives a hint of pinkish-red colour, which is typical of this material. When viewed by transmitted light, the material can appear sherry coloured from one angle and red from another.

Figure 1.17 Detail of colour swirls in burmite, showing spots of colour.

Figure 1.18 Piece of burmite, showing the prized burnt-orange to cherry-red colour (magnified).

- Burmite can contain many cracks, which tend to be very angular. They may appear white due to the intrusion of calcite.
- Although the largest piece of amber ever discovered came from Myanmar, it is usually found in small pieces. Carvings have on occasion been made of lots of little blocks of the material that have been stuck together.
- The much-sought-after clear burnt-orange to cherry red burmite usually occurs in smaller pieces than the other colours, and is more rare (Fig. 1.18).
- Under ultraviolet light freshly cut or polished burmite fluoresces strongly with a striking, mid-blue colour (Fig. 1.16). Less fresh pieces may fluoresce with a beige colour. Burmite also fluoresces blue in sunlight, and, like Mexican amber, this fluorescence appears to come from the body of the amber rather than from the surface.

Other ambers

Deposits of amber are still being found today, so any list can quickly be out of date. A relatively young amber was recently found in Borneo, and there are known deposits of Canadian amber, but this has been buried in the permafrost and is badly preserved. However, there are a few other types that should be mentioned and described as, though they are rare, they can be found in antique shops and seen in museums.

Chinese

Chinese amber is similar to burmite, and many of the old pieces of worked amber that have been labelled 'Chinese' may in fact be made of burmite. Some amber has been collected from near Fushan. It contains some insects and is believed to be between 60 and 35 million years old.

Rumanite

Rumanite came from the Carpathian Mountains in Romania. Both Tertiary and Cretacious deposits have been found. Although never plentiful, it was immensely popular about 100 years ago, but today is seldom seen except in museums and, on rare occasions, in antique jewellery. It occurs in shades of brown, and is very clear except for the mass of small fractures within the material. This is very typical of rumanite and seldom seen in other ambers.

Simetite

Simetite is also very rarely seen today. It is between 35 and 25 million years old and was found in Sicily near the River Simeto. It is predominantly orange to red in colour, sometimes appearing in a greenish hue. It is transparent, has few inclusions, and reflects light better than other ambers.

Cretaceous ambers

With the exception of burmite all the ambers mentioned so far are from the Tertiary period. There are many finds of amber from the Cretaceous period, 140 to 65 million years ago, but the vast majority of these are of little use as decorative material as they are far too brittle to work. They are usually found in very small pieces and tend to be a dull, opaque brown. They are, however, of great scientific interest, especially as some contain plant and animal inclusions.

Among the Cretaceous ambers worth mentioning are those from:

- **Japan**, which at 100 to 85 million years old is, together with burmite, the oldest amber from which objects have been carved. The largest deposits are north of Tokyo, near the town of Kuji.
- **USA**, where there have been several finds, for example in Alaska and Wyoming. The amber from New Jersey is aged about 90 million years old, and is of great interest as it contains a lot of insects.
- **Middle East**, especially Lebanon, which has brittle, transparent amber but which, at 125 million years old, is the oldest found to contain insects. Even more remarkable is the wide variety of insects found in it. It is from a weevil in Lebanese amber that the oldest DNA is said to have been extracted.
- **Siberia** has the world's largest Cretaceous amber deposit. It is thought to be about 100 million years old and lies in the permafrost, so is accessible for only a short period each year when the frozen ground thaws.
- **Europe** has its own Cretaceous amber. Deposits have been found in, among other places, England (the Isle of Wight), France, Denmark and Austria. They date from 130 to 100 million years old.

Treatment and uses of ambers

The following does not apply to Cretaceous amber, with the exception of burmite:

- Amber can be cut, carved, and turned on a lathe. It takes a good polish.
- Amber is usually used for jewellery though in previous years it was a popular medium for carving small figures. Today it is again possible to buy carvings – often small carvings depicting animals – in Baltic, Dominican and Mexican ambers (Fig. 1.12).
- Historically, ambers have been dissolved in solvents and used as varnishes, and occasionally used as incense.
- Poor-quality pieces of amber, and amber dust and chips from cutting, are still used today to produce amber oil for varnish or for medicinal purposes.

- There is at present research taking place into the possibilities opened up by amber. For example, the DNA from ancient bugs is still being sought, and the trapped gases and liquids are being investigated. Every type of amber is capable of giving us some information about our planet, for example the climate, as it was millions of years ago.

Treatments specific to Baltic amber

Baltic amber is treated and used in a variety of ways that are not suitable for other types of amber.

It darkens with exposure to the ultraviolet rays in sunlight, and the surface becomes opaque due to oxidation. Examples of beads carved in the Stone Age are usually a deep, dull, reddish colour, but when viewed by transmitted light they glow a deep brownish-red. If one of these beads were broken the inside would be much paler and probably clear. The length of time taken for this to happen varies according to where and how the amber has been stored. It can take from several years to several hundred years to occur naturally, but the effect can be simulated through treatment.

The following apply only to Baltic amber:

- Baltic amber goes through a process of cutting and several polishes with various grades of abrasives, before it attains its bright, shiny surface. It is water cooled during some of the processes to prevent damage to the surface from overheating.
- The most common use of Baltic amber is as beads. Tastes vary from country to country, some liking pale yellow, clear beads, others preferring deep golden beads. The evenly coloured, clear beads have usually been treated to clarify them. This can be done by heating them in an autoclave, usually in the presence of nitrogen, which has the effect of 'melting' the air bubbles in the material and of darkening the surface of the amber (Figs 1.19 and 1.32).
- Amber can also be darkened simply by heating it in an oven, which oxidises the surface but does not penetrate the material. The result is an opaque finish, giving the amber an older patina (Fig. 1.20).
- After heat treatment the amber is polished again. This must be done carefully as too much polishing will remove the darkened or opaque surface.
- Following the clarifying process the amber must be cooled at a controlled rate. It is at this time that the 'sun spangles' or 'lily pads' appear that have become popular and are generally considered proof of genuine amber. They are in fact discoidal stress cracks arising from the cooling process, which appear as irregular circles of varying size and which contain radiating lines. They reflect light very effectively. They can appear in untreated amber, but this is rare (Figs 1.21 and 1.22).

- The surface of amber can be dyed. It is now possible to buy amber in almost any colour, though to date this form of treated material – with the possible exception of red – has not gained much popularity.
- A relatively new treatment that is popular is given to cabochon cut amber for mounting in jewellery. It is first heat treated and polished in the normal way, and then burnt on the back. This gives the optical illusion of the amber being green or almost black. When viewed from the side only the base is dark while the rest is clear golden (Figs 1.25–1.27). If this burnt area is partly polished away, the amber can look red.
- The same effect of green or black amber is sometimes obtained by making amber doublets. This is done by covering the back of a cabochon with a sliver of black plastic.
- In past times amber has been made into a variety of objects, from drinking vessels to chandeliers. Much of the work was done on a lathe.

Figure 1.19 Baltic amber cabachons, showing the effects of progressive degrees of heat treatment. The palest bead is untreated.

Figure 1.20 Baltic amber, broken bead showing surface effect of heat treatment.

- Baltic amber was also used as inlay in furniture, small caskets or wall panels. A very small amount of this work is still done today. Thin slices of amber are applied as mosaics of different opacities and colours, sometimes with little pictures carved into the backs of the sections (called 'intaglio' work), or with a metal backing on which a design or picture has been etched (called 'eglomisé' work). If there is no metal backing, the amber is laid over wood that has been coated with a white substance (or, occasionally, with gold leaf), in order to show up the colour of the amber. Mounts, edges and locks are usually of ivory as it is less susceptible to damage if knocked, and the two materials complement each other aesthetically (Fig. 1.39).

Pressed and reconstituted Baltic amber

Baltic amber can be reconstituted in various ways, with or without the addition of artificial fillers and dyes.

- **Polybern** is a material that uses amber chips submerged in a synthetic polymer such as polyester or acrylic to which an amber-coloured dye has been added. The name 'polybern' is made up from the words polymer and the German word for amber, 'bernstein'. The product is immediately recognisable as the chips sink to the bottom of the material and are clearly visible. They tend to have sharp edges and appear as lumps rather than as the gentle variations in colour and opacity of natural amber (Figs 1.28 and 1.29).
- Small chips of amber can be pressed together under heat and pressure to form a larger solid mass, without the addition of synthetic resin. This **pressed amber** can be cut and polished as natural amber. During the process the edges of the amber chips oxidise and darken, so that when viewed by transmitted light it is possible to see dark, fuzzy outlines of all the pieces of amber (Fig. 1.30). This material should not be confused with 'polybern'.
- A more convincing method of making **pressed amber** is by melting small chips or powder under heat and pressure, also without additives, then extruding it through apertures to form a rod for cutting and polishing. The resulting material produces elongated swirls of clear and opaque material, often with a feathered pattern (Fig. 1.31), and is sometimes called '**ambroid**'. Dyes can be added in this process, green or red being the most common colours. This material was extensively used for smoking requisites in the nineteenth century. A very clear version was also made, and used for jewellery. This is called simply 'pressed amber' (Fig. 1.32). Today it is still being made and sold through less reputable dealers as natural amber.

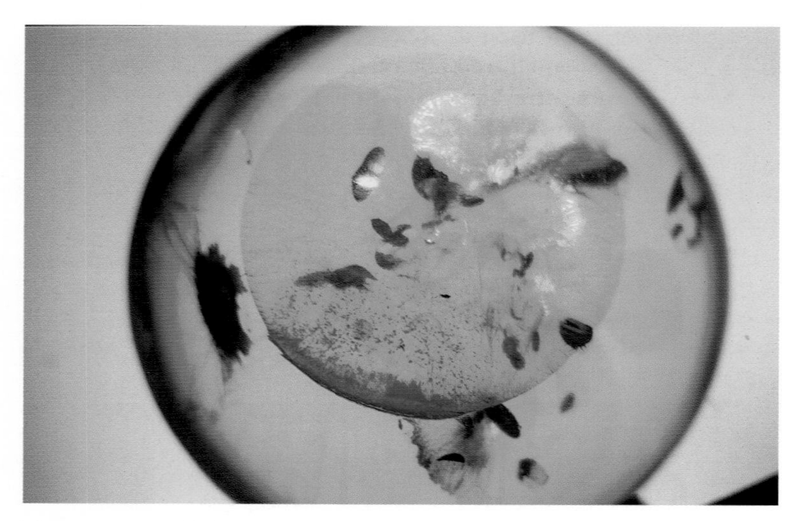

Figure 1.21 Baltic amber cabachon showing with discoidal stress fracture (magnified).

Figure 1.22 Detail of discoidal stress fracture in Baltic amber.

Figure 1.23 Plastic simulant with imitation stress fractures.

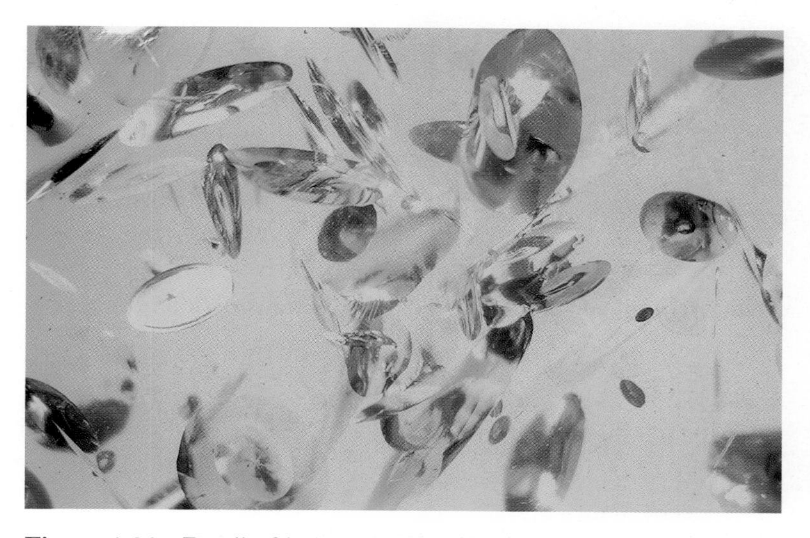

Figure 1.24 Detail of imitation stress fractures.

Figure 1.25 Heat-treated 'green' Baltic amber cabochon.

Figure 1.26 'Green' amber cabachon showing burnt surface on the back.

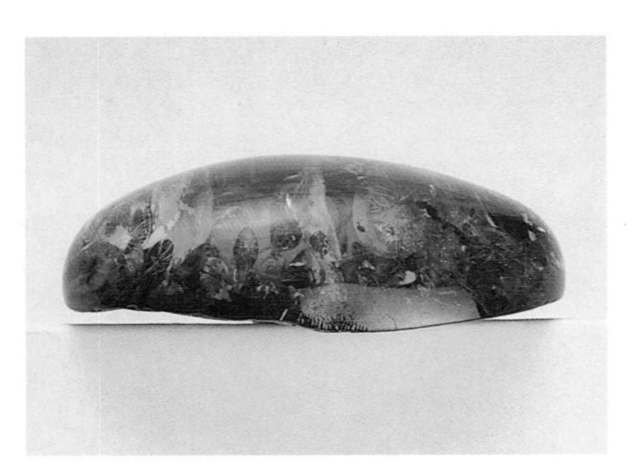

Figure 1.27 'Green' amber cabachon, side view, showing golden colour.

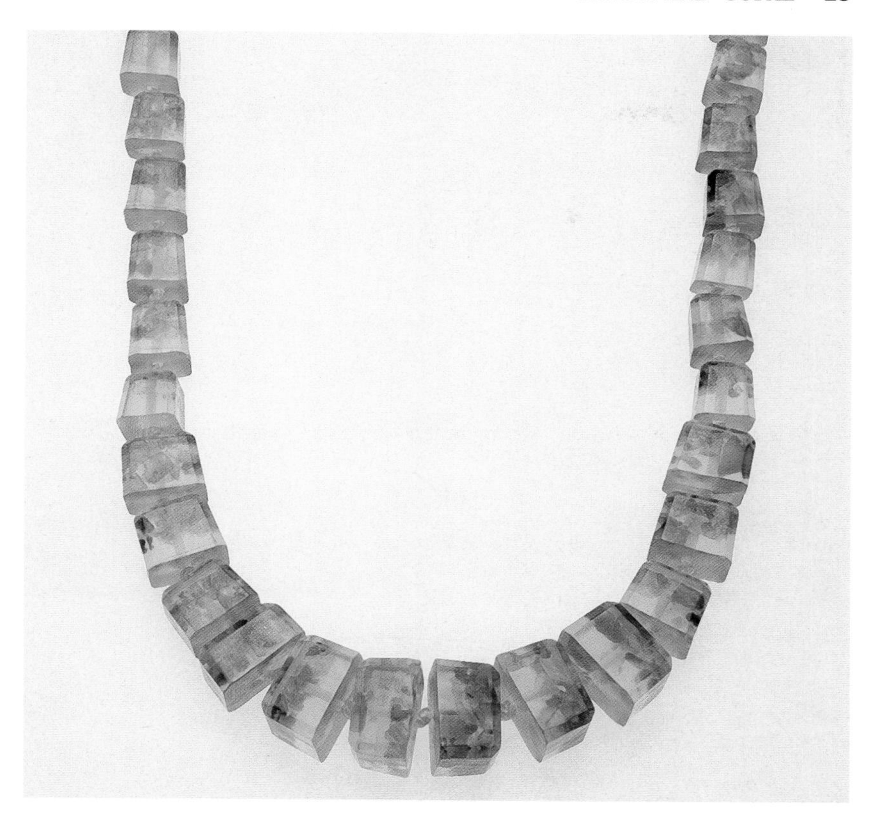

Figure 1.28 Necklace of 'polybern' beads.

Figure 1.29 Detail of polybern disc.

Figure 1.30 Pressed amber, showing oxidised outlines of amber pieces.

Figure 1.31 'Ambroid' pressed amber, showing 'feathering'.

Figure 1.32 Old pressed amber beads (Victorian), and heat-treated Baltic amber beads (modern).

Figure 1.33 Detail of old pressed amber beads, showing surface deterioration.

SIMULANTS

- **Reconstituted ambers.** See the previous section.
- **Plastic** is the most common simulant for amber, and it has been used since the early plastics came into production. Phenol formaldehyde – which is better known as Bakelite – was especially successful. It was dyed either a golden colour to resemble Baltic amber, or red to resemble burmite. Nowadays it is quite common to come across old jewellery made of Bakelite (Figs 13.4 and 13.5).

  In the middle of the twentieth century carvings were coming onto the market that were made of Bakelite and were intended to imitate old Chinese carvings in burmite. Visually they were very successful and for a while even some museums were fooled, but tests proved them to be fakes.

  Viewed by transmitted light, the colour of Bakelite appears too even (Fig. 13.6). (Further tests are listed later in this chapter under 'Tests'.).
- **Modern plastics** are today used to fake amber. To give the impression of sun spangles, tiny slivers of metal are added to the plastic. When examined under a 10× lens these are seen to be too regular, often too thick, and without any radiating stress lines. They are often more easily discernible as fakes when they are photographed. Modern plastics are also used to fake pieces of amber with insect inclusions. The 'amber' is often very clear and the insect beautifully placed in the middle of the piece. The insect normally proves to be of an extant species as opposed to an extinct one (Fig. 13.7).
- **Doublets.** More worrying and harder to discern are the faked insect inclusions that have been carefully placed on a piece of amber and capped with plastic. Careful inspection will probably show joins, and the piece is likely to sink in salt water because it has a higher specific gravity (a measurement of the mass of a substance) than natural amber (see 'Tests'). Again, the insect species will most likely be extant.
- **Faked inclusions.** Most worrying of all are the insect inclusions faked in natural amber. Here a piece of amber is cut into two or three pieces, small sections are scooped out and insects inserted into them, possibly in melted copal resin to give the correct characteristics in testing. The whole thing is glued together again with copal. This can be done so cleverly that the joins look like natural cracks or swirls in the material. Examination under a microscope may show more, but it is sometimes necessary to ask an entomologist to identify the insect species in order to be certain. As with the Bakelite carvings, even museums have been fooled by these fakes.

- **Copal** is sometimes sold as amber and can be difficult to distinguish, especially for people not used to handling the materials. It is also used as filler or glue in the manufacture of faked amber inclusions, as it has the same colour and specific gravity as amber, and is therefore less easily detected than other materials.

TESTS AND IDENTIFICATION OF AMBERS AND COPALS

As with all organics, the safest method of identification is by sight and feel, as almost all tests are destructive to some degree. Even so, it can be necessary to carry out careful testing as amber is notoriously hard to identify. Having once been a liquid, natural amber can and often does contain swirls of colour and air bubbles, which are the signs that are usually indicative of plastic imitations.

Visual examination

- Amber and copal feel light. This is the most useful test when beachcombing for amber, which looks dull and pebble-like in its rough state. In other situations it must be remembered that plastic is also light.
- Amber and copal should feel warm to the touch – never cold. This also applies to plastic.
- Amber and copal break with a conchoidal fracture (Fig. 1.7).
- Evenness of colour viewed by transmitted light in a slightly opaque piece may indicate plastic, especially if it is coloured red (Fig. 13.6).
- A piece that is very even in colour and totally clear may indicate plastic. Although it does occur, totally clear natural amber that is free of any form of colour variation, crack or inclusion is unusual.
- Spots of colour may indicate plastic, though some natural resins, for example burmite, have swirls of colour composed of microscopic spots.
- Discoidal stress fractures ('sun spangles') in natural amber that has been heat treated are uneven and have lines radiating from the centre of each spangle (Figs 1.21 and 1.22). Viewed side-on they become almost invisible. Artificial sun spangles in plastic tend to be even in shape and when viewed side-on they remain obvious (Figs 1.23 and 1.24).
- Unlike other gem materials, it is common to see swirls and bubbles in amber and copal as the material does not form in the same way as mineral crystals. It is occasionally possible, though rare, to see double bubbles of gas and liquid inside these resins.
- Natural amber can have inclusions of plant debris, dirt, small animals and minerals.

- Lumps or straight lines in the colour variations can indicate amber chips in plastic ('polybern') (Fig. 1.29).
- Fuzzy, blurred lumps of darker colour in a pale matrix indicate pressed amber (Fig. 1.30).
- Lines of opaque and clear material and feathered swirls indicate pressed amber ('ambroid') (Fig. 1.31).
- Surface deterioration displaying crazing may indicate very old, oxidised amber. The presence of white powder with the crazing indicates copal (Fig. 1.35).
- Surface crazing that is clear, on items of amber jewellery of even colour – either transparent or opaque – could indicate pressed amber from the end of the nineteenth century. Typical examples are beads, or brooches with a motif of flowers or fruits in clear material, on an opaque background (Fig. 1.33).
- Pale, scratched areas on dark amber indicate surface treatment, for example heat treatment which has oxidized the surface, or a coating of dye.

INSECT INCLUSIONS

- Completely clear pieces with well-placed insects indicate fakes. Most insects will have struggled to escape, leaving the resin disturbed, and possibly breaking legs or wings in the process. More resin subsequently covering the insect would in all probability crush it further. A perfectly displayed specimen is therefore suspect (Fig. 13.7).
- Cracks or joins around the insects can be due to another layer of resin covering it naturally, but this usually produces wavy lines. Straight lines are especially suspect.
- Unusual or too large insects should arouse suspicion. Obvious fakes include tiny fish or other marine life, or anything too heavy or strong to have become stuck.

Tests

NON-DESTRUCTIVE TESTS

- **Saturated salt solution.** Amber has a specific gravity that is less than that of salt water, so it will float. It is easy to test items by immersing them in saturated salt water (the water does not need to be heated). This test is useful for testing large pieces such as carvings, but cannot be used on jewellery that is set as the metal will give an incorrect result. When testing beads that are strung, it should be remembered that the thread may alter the result. The air in an unstrung drilled bead may help to keep it afloat, so it is important to ensure that no air is trapped in the drill hole. Plastics used to imitate amber sink in salt water due to their higher specific

Figure 1.34 Pieces of rough and polished copal. Top, left and middle: South American copal; right: New Zealand kauri; bottom right, Dominican copal.

Figure 1.35 Detail of copal surface, showing deterioration and 'dandruff'.

gravity. Any items tested in this way should be thoroughly rinsed in cold water afterwards as the salt solution can seep into small fissures leaving a white residue when it dries.

- **Ultraviolet light.** Most amber and copal fluoresces in UV light (Figs 1.13 and 1.16). This fluorescence dims if the amber has been treated with heat to clarify it or to press it. It also tends to dim and change colour with age – freshly cut or polished amber will always fluoresce more strongly.

  Some ambers fluoresce in daylight, possibly with a different colour to that displayed under ultraviolet light, and sometimes the fluorescence comes from the body of the amber, rather than from the surface. This is often noticeable in light that is strong in the blue wavelengths. Burmite and Mexican ambers fluoresce in daylight, but Baltic amber does not. It should be remembered that daylight – even sunlight – varies greatly in colour, so the fluorescent effects of ambers can be difficult to judge in this light.

  Most plastics do not fluoresce, either under UV light or in daylight. The exception is casein which fluoresces under UV light.
- **Electric charge.** Rubbing a piece of amber or copal with a piece of cloth or fur produces a negative electric charge, strong enough to pick up items such as small feathers. This gave rise to the Greek name for amber, 'elektron'. This test is inconclusive, though, as many plastics behave in the same way.

Destructive tests

- **Sectility.** When pared with a sharp knife, amber and copal will chip, while plastics will pare. Only a tiny scraping should be taken from the back of an item, or from a drill hole in a bead.
- **Solvents.** A small drop of solvent such as acetone or methylated spirits on the back of the item to be tested should not, in theory, affect amber, but it will turn copal sticky. Depending on the age of the resin, the test may take some minutes (younger resins dissolve faster) and more than one application of solvent may be needed, as it evaporates quickly in air. It should be noted that solvents can attack many plastics and some ambers, so it is not a conclusive test. Also, care must be taken to use only a small amount of solvent, as too much may cause extended damage.
- **Burning.*** Copals melt very fast (at 150°C) and give off a delightful, resinous smell. Ambers burn rather than melt, at a higher temperature. Some give a resinous smell, but Baltic amber has a very characteristic – sometimes described as 'soapy' – and less attractive aroma due to its succinic acid content. Most plastics smell acrid and unpleasant when burned. This test is best carried out by taking a tiny scraping of the material and heating it on a knife blade.

- **Hot point.*** Similar to the above test, a needle is heated to red hot in a flame and touched against the surface of the item to be tested. It will make a small indent in a piece of amber, but will pass quickly into a piece of copal – the younger the resin the faster it melts. It should be noted that this can leave a very large hole in some materials. The hot needle may also affect plastics.
- **Laboratory tests.** There are one or two further tests that can be carried out in a laboratory using specialised equipment. The test most commonly used for amber is infrared spectroscopy, which detects the chemical components of a given specimen. It can, for example, discern the differences between ambers from different localities, and is especially good at distinguishing between succinites and retinites (those ambers that contain succinic acid and those that contain none). It can also distinguish between various man-made polymers.

*Note: Care must be taken with the burning and hot-point tests as one of the early plastics (celluloid) *will ignite* when heated. It is marginally less dangerous to heat a scraping of the material to be tested than to stick a hot needle into the whole item.

Copals

As with amber, copal deposits are found worldwide in subtropical regions, for example Brazil, the Dominican Republic, Columbia, Australia, New Zealand, West and East Africa, and Madagascar, to name but a few. Much of that found is probably no more than a few hundred years old, though some is considerably older.

The oldest known copal is found in Japan and is reckoned to be about 33 000 years old. Its characteristics are now midway between those of copal and those of amber as it is partially cross-linked and polymerised.

Most copals come from leguminous trees, especially the genus *Hymenaea*, which has also produced some ambers. Those from West Africa come from the genus *Copaifera*.

Copal is generally paler than amber, and is often the pale yellow colour of dry sherry (Fig. 1.34). Having been produced in the same way as amber it, too, can contain lots of flora and fauna inclusions. Copal is less stable than amber and the surface deteriorates relatively quickly, displaying the typical crazing which produces a whitish powder, sometimes referred to as 'dandruff' (Fig. 1.35). The surface of amber can also craze with time, but the pattern is darker and does not produce white powder.

Copals fluoresce under ultraviolet light, with a chalky yellow colour, sometimes with a hint of pale blue. They do not fluoresce in daylight.

Figure 1.36 Copal seeping on trunk of Kauri tree.

New Zealand kauri

The most famous of copals, the so-called kauri gum, is from the *Agathis australis* (Fig. 1.36). Examples of this tree can still be seen today in a couple of the national forests in North Island, New Zealand. The species almost died out following overcollection of the gum. The problem was exacerbated by the popularity of the wood, which, growing very straight and tall and with no lateral branches but only a crown at the top, was ideal for everything from ships' masts to house building. Today all collection of copal, which is gathered from around the roots of the trees, is regulated, and indeed is only allowed if a tree falls from old age or other natural causes. Use of the wood is regulated in the same way.

Kauri gum is of a darker colour than most copals (Fig. 1.34), and, like the Japanese copal, some of it is turning into amber. It is mostly transparent and it contains few inclusions. Insect inclusions are extremely rare. It has been successfully worked into a variety of decorative objects and jewellery.

Dominican Republic

This copal, like the local amber, is from the genus *Hymenaea*. It is the typical pale yellow colour of most copals, and is extremely rich in plant and animal inclusions which are beautifully preserved and very distinct in the clear, pale material. It is usually sold cut and polished to show off these inclusions (Figs 1.34 and 1.3).

As with most copals, it is not very stable and will deteriorate with age. For this reason it is regrettable that much is sold locally as genuine amber, even though it is fascinating in its own right. It is in plentiful supply.

Other copals

There is a lot of copal on the market from Madagascar. Its age is not known but it is rich in inclusions, especially flat-footed beetles.

As previously stated, there are many other copals found today, and it is hard to distinguish one from another without specialised testing such as infrared spectroscopy. The copals typically have the same pale yellow colour and rough, crazed surface. Although this surface is often removed before the material is polished and sold these copals are seldom worked to any great extent and is sold more as a curiosity.

Treatments and uses of copals

Copal does not undergo treatments to clarify, colour or darken it, as does Baltic amber.

- Copal can be cut, polished and made into jewellery, especially bead necklaces. It must be worked with caution as, having a relatively low melting point, the friction generated in cutting and polishing can damage the surface.
- Copal can be carved into figures and polished. However, copal deteriorates with age and jewellery or carvings lose their surface polish. There are carvings in existence that are many years old and do not yet show signs of deterioration, but these are made from old copal such as kauri.
- Though not as attractive as the more famous natural resins, frankincense and myrrh, copal gives a pleasant smell when burnt as incense.
- Copal, along with several other plant resins, can be powdered and melted for uses such as varnish, adhesive and in the manufacture of linoleum. Copal is especially good as a varnish for oil paintings or musical string instruments.

Simulants

Copal is plentiful and is not copied, but it is sometimes used and sold as a simulant of amber.

Conservation status

There is no ban on mining the various amber types, but it is a strictly regulated process in most countries. However, politically sensitive or troubled areas make regulation difficult. For example, it is only recently that mining of amber in Myanmar has been resumed.

Among the copals, only New Zealand kauri is totally protected. Although it is possible to walk in the national forests where a few of the trees still grow, it is requested that the trees are not touched as this would involve treading on the delicate, shallow root system. The trees may not be felled but should a tree fall from natural causes such as a storm, both the wood and any copal found around its roots can be collected by the appropriate authorities and sold. It is thus still possible to buy small amounts of kauri copal in New Zealand.

Past and present uses

Amber

Few materials carry more myths and legends than amber and it is not hard to understand why it has been so prized for over 10 000 years.

It is warm to the touch even when picked up in the sea, it catches the light and glows with sunshine colours, and when rubbed it has the ability to lift light objects such as feathers. Surely, it must have magical qualities!

It is known that Christopher Columbus saw Dominican amber being mined and traded, and amber carvings have been found in grave sites in the Americas. But both the graves and Columbus's voyage are relatively recent in terms of amber history. The use of Baltic amber has been traced back to the Stone Age.

In 7000 BC, around the area today called Denmark, man already knew how to cut and polish amber. Using a piece of bone or a stone he could chip it into shape, and polish it up with a bit of leather or fur. He made talismans, engraved simple patterns on them and bored a hole at one end so that they could be worn on a string of some sort.

Later his descendants were carving simple figures of animals that they saw, such as birds and bears (Fig. 1.37). At the same time they made lots of beads. Some of these have been found in graves where they had been used to adorn the deceased, but huge hoards of them have been unearthed from peat bogs where they were placed in clay pots, evidently as an offer-

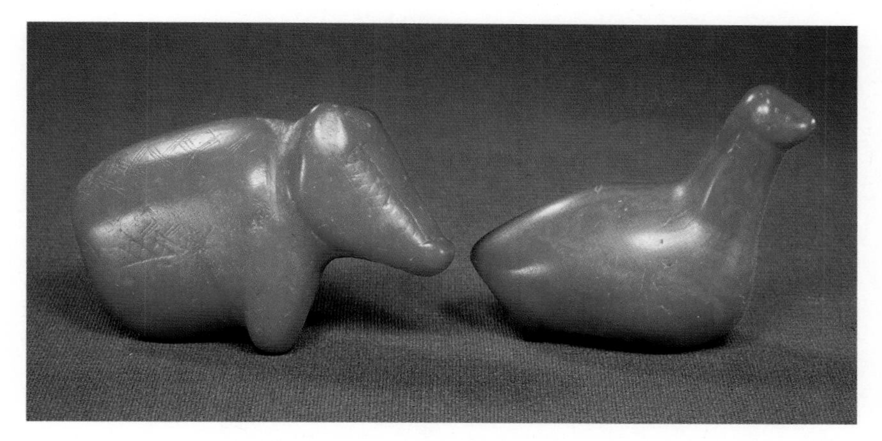

Figure 1.37 Bear and web-footed bird. Found separately in Jutland, Denmark. Dated 6800–4000 BC. *The National Museum of Denmark.*

ing to the gods. The objects found in the bogs are much better preserved than those found in earth, and though mostly dark red-brown, they still display some clarity. The finds dug out of graves resemble compressed drinking chocolate powder and have lost all their former glory.

Archaeological digs have shown that by the Bronze Age 3000 years later, there was less amber around in the north, but instead there was copper. This was imported together with salt, and traded for amber which was becoming popular further south. It has been found, for example, in Greek graves dating from this time.

The Romans loved amber. They imported large quantities of rough material, which they cut and polished and even turned on lathes in their own workshops on the Adriatic coast. Wearing amber was strictly a sign of status, and its use was regulated. Documentation exists that it was worn by gladiators, and there is also a suggestion that it was used to adorn horses on ceremonial occasions.

By the third and fourth centuries AD amber was again being worn in the north, and the Vikings had their own workshops. During all this time it was a sign of wealth, and was also believed to bring luck.

Little was used in the Middle Ages, either at home or as an export material, but it became popular again in Europe in the fourteenth century, when there was a great demand for rosaries. Collection of amber from the eastern shores of the Baltic was strictly regulated and anything found belonged to the local duke. Stealing was instantly punishable by hanging. The rough material was sent to the cities for working, and amber guilds developed in the cities. The guilds kept tight control on the whole business of working and selling the material. The most famous of the amber cities and the one that probably produced the most beautiful work was Königsberg.

Figure 1.38 Chinese carved buckle. Baltic amber. Qing dynasty.

In 1899 control of amber was taken over by the Prussian state. By this time it was no longer dredged from the shoreline but mined in open pits after the topsoil had been removed. It is still mined in those areas today.

The good-quality amber was used for decorative purposes while the off-cuts and poor-quality material was boiled down in oil, thinned and used as a hard varnish for wood, especially for musical string instruments or for ships.

The European Golden Age of amber work was in the seventeenth and eighteenth centuries when master carvers, often foreign, were employed by the royal courts to work in amber, ivory and precious metals. Even royalty themselves dabbled in working the materials. An example was the Danish royal court, which was then very wealthy. As a result the museums in Denmark contain fine examples from the period, including whole chandeliers in amber and a set of dinner plates made of amber and silver. Most famous of their carvers was probably Lorentz Spengler, who was born in Germany, moved to England for a time, and then on to Denmark.

The work done at this time was very varied, from carved figures to eglomisé work and marquetry. Items produced included such objects as table ornaments, caskets and tankards. Much of the amber work was mounted in silver and combined with ivory (Fig. 1.39).

The most fantastic and fanciful example of this type of work has to be the famous Amber Room built in 1711 for Frederick I of Prussia. The room consisted of wooden panels inlaid with amber mosaics, some of them backed with gold leaf to emphasise the colour of the amber. Mirrors and furniture for the room were similarly adorned, with the result that pretty much every surface of the room was covered. The effect, if not stunning, must certainly have been startling. Originally installed in Berlin Castle, the room was later dismantled and moved to the Winter Palace in St Petersburg as a present for Peter the Great. In 1755 his daughter had the room dismantled again and moved to the

Figure 1.39 Amber and ivory casket, made by Gottfried Wolfram, 1707. *Rosenborg Castle, Copenhagen, Denmark.*

summer residence of the Tsar and his family at Tsarskoe Selo. Here it stayed until the Second World War. When the Germans attacked in 1941 the amber panels and furniture were dismantled and taken away. To date no one knows if it was looted and broken up, burnt in a warehouse fire, or hidden where nobody could find it. It remains one of the mysteries of the twentieth century. Recently amber craftsmen have slowly recreated the panels and furniture from old photographs, and the room is rebuilt at the Summer Palace.

By the nineteenth century amber was again being used as jewellery. Necklaces, bracelets, rings and other items were being made. Some simple boxes and caskets were made but the intricate work of the Golden Age was not repeated.

The late nineteenth century saw the appearance of pressed amber, which was popular for smoking requisites and jewellery. Favourites pieces were faceted beads and brooches depicting flowers and fruit. Early plastics also appeared on the market and pretty, red Bakelite beads imitated burmite. When the film *Jurassic Park* came out at the end of the last century natural amber saw a renaissance, and it remains popular today.

Amber off-cuts and dust are still used for varnish, and for alternative medicine products, where it is turned into creams to ease aching joints and muscles. It has long been thought to help these problems, though 200 years ago a whole lump of the material was placed on the afflicted area rather than using an ointment.

The Romans were great believers in the medicinal powers of amber. They were convinced that it could cure coughs, ear-ache, poor eye-sight and fits, or when burned, the fumes could help women in labour. Amber was also thought to have magical powers, such as the ability to make a sleeping wife confess all her evil deeds when a piece was placed on her breast.

It is only recently that we have discovered the origins of amber. For millennia it was a puzzle, especially as it apparently came from the sea, so nobody imagined that it could have anything to do with trees.

There were many myths surrounding its origins. One myth from Poland told that there was a flood that lasted 40 days and nights. The people wept so much that their tears fell into the water and turned into amber. The clear amber was the tears of the innocents, the dark amber was the tears of the sinners, but the useless amber came from evil men.

Other beliefs were that it was the tears of birds, or moisture from the sun's rays as it set. Pliny the Elder (who lived in the first century AD) believed that it came from trees, though few people agreed with him at the time. However, he had no idea of its age and thought that it solidified through the effects of frost, or by having been submerged in sea water.

Copal

Our knowledge of the history of copal is also confined to one species: New Zealand kauri gum.

Fresh resin was originally used by the Maoris as chewing gum. They also burned it to keep away midges and used the blackened material as a dye for tattoos. It was first taken to Australia and England in the early nineteenth century. Travellers from those countries traded with the Maoris who had dug the copal from the ground. By the end of the century the trade had developed into an industry employing thousands of diggers, including many Europeans, and the copal was exported for use in the manufacture of varnishes, linoleum, sealing wax, candles and dressing for calico.

During the First World War the industry ceased, but it started up again after 1918, when digging was supplemented by tapping the trees for fresh resin as well. This habit, together with the felling of the trees for their beautiful wood (which had been used in ship and house building for over a century) decimated the kauri forests. The trees do not reach maturity until they are about 300 years old, and they can live for 2000 years. In only 150 years the forests were reduced to just 4 per cent of their original volume, and kauri gum and the beautiful trees that produced it are virtually consigned to history.

2 Jet

Jet is a fossilised wood. It is a coal but differs from other types in that it derives from a single species of tree, rather than from various trees and plants. Good-quality jet is found in small quantities in a few localities only, so it is relatively rare as a gem material.

Structure and properties

In times past jet has been extracted in such countries as Russia, Turkey, Germany and France as well as in England and Spain, but of these only England and Spain are still sources of gem quality material.

Today jet is also available from various other places, for example Utah and Colorado, New Mexico, China, and the Ukraine, but it is not of the same quality. Much of it is dull in appearance, and some, though it takes a high polish, is brittle and liable to crack when worked. It is possible that the difference in qualities lies in the age of the material. For example, Chinese jet is reckoned to be less than a quarter of the age of English jet, and though it can be carved, it cannot be polished to the same high lustre. The material from New Mexico tends to fracture (Fig. 2.1). To a purist, these materials are not true jet.

The following therefore refers to European jet, especially that from around Whitby in Yorkshire, England.

- Jet derives from trees related to the *Araucaria* family (which includes the monkey puzzle tree). It is found in the Jurassic, Upper Lias strata, and is about 180 million years old.
- The material developed in anaerobic conditions. The trees died and fell, possibly ending in swamps or rivers and being swept down towards the sea. As they started to rot and break up they sank to the bottom and became embedded in the seabed. Over millions of years, layers of other decaying organisms, plant and animal, piled up on top of the wood, burying it and causing heat to develop. The

Figure 2.1 Various jets, clockwise from bottom: Whitby half polished on top of Whitby rough, Mexican half polished, Mexican rough, Ukranian rough, Chinese polished.

lack of air prevented further decay. Later, geological changes caused the seabed to lift above sea level so that we now find jet on land.

- Jet occurs in small seams of around 20–25 cm thick and less than a metre long.
- Whitby jet occurs in both 'hard' and 'soft' qualities. The latter is unsuitable for working as it is slightly brittle and may at some time crack. Although the reason for the differences is uncertain, there are theories that soft jet may have been formed in fresh water, or that it may be younger. It is also possible that soft jet contains more sulphur, which prevents the material from hardening in the same way. A jet worker can see immediately which material he is dealing with, though people with less experience may find it difficult.
- Whitby jet also occurs as 'plank' or 'core' jet. This refers to the way it has been formed: either completely flattened or with a silicified core.
- There is little structure to be seen in jet, though in rare specimens of rough material it is possible to see the annual growth rings of the original tree, which have been compressed to flat lines.
- When found on a beach jet has a dull, water-worn appearance (Fig.

Figure 2.2 Piece of Whitby jet showing conchoidal fracture.

2.1). When dug out of the ground it is covered in a greyish skin called 'spar'.

- Jet displays a conchoidal fracture when broken (Fig. 2.2). The break shows the material's deep, velvety black colour and high lustre. Polishing also brings out the colour and lustre (Figs 2.1 and 2.14).
- Although sufficiently hard to cut, turn and polish to a high lustre, jet is relatively soft. This means that, though its lustre does not diminish with age, it can become scratched. The colour remains unchanged and does not fade.
- When worked, jet produces a brown dust, and when rubbed across a rough surface will leave a mid-brown streak of colour.
- Jet is a poor conductor of heat. It feels warm to the touch.
- Like amber, jet can become sufficiently electrically charged when rubbed to be able to pick up items such as small feathers.
- Also like amber, jet is very light.
- Chemically, jet is about 75 per cent carbon with some hydrogen, oxygen and sulphur. There can also be traces of many other minerals.

Treatments and uses

Jet is used mostly for jewellery and small, decorative items. It is best known for its use as mourning jewellery in the nineteenth century.

Figure 2.3 Jet and jet simulants. Top row: vulcanite, jet; bottom row: vulcanite, bog oak, jet.

- Jet can be cut, carved, turned on a lathe, and polished to an almost vitreous lustre (Fig. 2.3).
- Jet cannot be softened or moulded.
- Jet is neither dyed nor bleached.
- To make up larger items, pieces of jet are glued together. Screws are never used. The traditional glue used by the jet workers in the nineteenth century was an animal derivative called 'ockamatutt' (Fig. 2.11). Today epoxy glue has taken its place.
- Due to its colour, jet is ideal for use in combination with other materials such as ivory, for example as cameo doublets, with a jet back topped by an ivory carving, or as triplets of jet, ivory and jet. It has also been used to frame painted miniatures or shell cameos.
- Jet has been used in mosaics, for example the Florentine mosaics called 'pietra dura', which combined coloured stones.
- In the past jet jewellery was seldom mounted with a metal surround. Today it is mostly mounted in silver.
- Late nineteenth century necklaces were sometimes made of beads strung on thread in the traditional way, but also very typical were those made of carved jet links (Fig. 2.14).

SIMULANTS

There exists a wealth of old, jet simulants. Several materials used to make jewellery were black, and further, almost anything could be dyed black. It was much harder to produce plastics in pale colours, or to bleach materials, to imitate a material such as ivory. There are also other coals which were worked in their own right and not necessarily intended as a simulant of jet, but which may be confused with jet.

- **Cannel coal** is a dull, compact, bituminous coal, that is to say it is rich in volatile hydrocarbons and burns very easily. It consists of varied plant debris, and under magnification it can be seen that it is made up of spores. It is brittle with a slightly paler sheen than that of jet and does not take such a high polish. It has a conchoidal fracture but produces a black powder when worked. It was more commonly used for large pieces than for jewellery. It leaves a black streak when rubbed against a matt surface such as unglazed porcelain.
- **Kimmeridge shale** is a bituminous shale which resembles jet quite closely. Many objects found in burial sites were made of this material. It is light and varies in colour from black to brownish or greyish black. Under magnification it may be seen to contain marine fossils. Although kimmeridge shale takes a high polish, unlike jet it dulls with age and tends to crack.
- **Anthracite** is very black, but it is brittle and has a glassy appearance. It can be successfully made into small items and was used in America for this purpose at the turn of the twentieth century. It gives a black streak.
- **Bog oak** is from Ireland. It is paler in colour than jet and has a matt surface (Figs 2.3 and 2.5). It is much younger than jet and is a semi-fossilised wood found in peat bogs. Under magnification its woody structure is visible. Bog oak is tougher than jet and does not display a conchoidal fracture. Although it had been carved locally for a very long time before the Victorian vogue for black jewellery, it followed the popularity of jet as mourning jewellery and became a cheaper alternative. Carvings were often of a local theme such as a shamrock or a harp.
- **Horn** was a common simulant of jet, and can sometimes be found mixed with jet in items such as necklaces. It is light and easily manufactured as it is thermoplastic and could be moulded into shape (see Chapter 7, 'Horn'). When dyed black it makes a convincing imitation though its slight translucency around the edges of a carving, where the material is thinner, may be detected by transmitted light. It lacks the deep, rich colour and high lustre of jet (Fig. 2.7).

- **French jet** is glass. It was first made to satisfy the increasing demand for black jewellery in France and was cheaper than jet as it could be mass produced. Later it was also made in England and called 'Vauxhall glass' (Fig. 2.4). It may show some transparency by transmitted light – usually a dark red colour due to its manganese oxide content. It is much heavier than jet or any of the other simulants, and is cold to the touch. It has a bright, glassy lustre and can contain air bubbles or display marks from moulding. Being glass, French jet has a conchoidal fracture. It is usual to see lots of small fractures where it has shattered, rather than a single break.

Figure 2.4 'French jet' glass beads.

Figure 2.5 Detail of bog oak, with typical motif.

Figure 2.6 Jet simulant showing uneven colour.

The following are all man-made polymers that were used in the early to mid-twentieth century, and resemble jet. They may be encountered in the antiques world today. They also appear in Chapter 13 ('Plastics') where they are dealt with in more detail.

- **Vulcanite** is by far the most common of the jet simulants. It is also known as 'ebonite' and is regarded as an early plastic. It was made by vulcanising rubber, that is to say by heating rubber with sulphur. It could be moulded and mass produced and was therefore much cheaper than jet. It was totally opaque and colour was added to make it a deep black. The colour faded with age and today most pieces display a greenish-brown hue (Fig. 2.3). They have also lost their high polish. Under magnification tiny spots of varying colour may be seen, as can a slightly pitted surface (Fig. 2.6). Vulcanite did not split when worked so items of jewellery could be screwed together (Fig. 2.11). The material was not completely rigid and, in the case of necklaces, the links could be bent open to be joined (Fig. 2.9). Like jet, it gives a brown streak.
- **Gutta percha** in its pure form is white, and in crude form can be brown or black. It has the same chemical formula as rubber but a different physical state. It was not vulcanised as it was far too valuable a material and was prized for its elasticity and plasticity, and would have been ruined by vulcanisation. It was very easy to mould and could take intricate patterns, but to be made into jewellery, it would have needed to be hardened in some way. Gutta percha was less durable than vulcanite, and cracked and deteriorated when exposed to air over a long period of time. Instead it kept

well in cold, dark and wet conditions, which is why it was much used for covering underwater telegraph cables. Even so, there are many that believe it was widely used for jewellery. This confusion possibly arises as the name 'gutta percha' became a generic term used for many materials.

Figure 2.7 Horn chain link, cut only on one side, and showing striated structure.

Figure 2.8 Jet chain link which has been cut through.

Figure 2.9 Vulcanite chain link which has not been cut through.

- **Bois durci** is often mentioned as a simulant for jet but was in fact seldom used because it was formed in large, steel moulds, which were not suitable for the production of small items. It was a mixture of hardwood flour and ox blood, coloured with lampblack (a form of soot). It took a good, detailed imprint, was durable, and had its own niche, being used for plaques, barometer cases, inkwells and other large items (Fig. 13.2). The material gives a brown streak, but the words 'Bois Durci' often appear moulded into the back or bottom of an object.
- **Shellac** was a substance excreted by insects. It was mixed with wood flour, moulded and dyed – usually black or brown. It was lighter than bois durci and gave good detail in moulding. It was used to make picture frames (Fig. 13.1), dressing table sets, some jewellery and, notably, 78 rpm gramophone records. Examined under a microscope it is possible to see the wood content. Black shellac gives a black streak.
- **Celluloid** was one of the early plastics that was used in numerous ways. It was made of cellulose nitrate and was extremely combustible, but a safer version, cellulose acetate, was later developed. Dyed black, they are reasonable imitations of jet but they have a plasticy look and feel. Black celluloid gives a faint, black streak.
- **Bakelite**, or phenol formaldehyde, could be dyed any colour and was widely used. When dyed black, it was also a reasonable jet imitation, but it dulls with age and has a plasticky look and feel. Black Bakelite gives a black streak.

TESTS AND IDENTIFICATION

It is usually sufficient to examine an item with the naked eye and with a 10× lens to determine whether it is jet or one of its simulants. Further tests can be carried out, but most are destructive.

Visual examination

- Seen with the naked eye, jet has a deep, rich, velvety black colour – the blackest of blacks – and it takes a mirror-like polish (Figs 2.2 and 2.14).
- It is very light, and warm to the touch.
- Viewed under magnification, it is sometimes possible to see signs of flattened annual growth rings in the wood.
- Jet is always carved and cannot be moulded, so signs of carving and cutting should be visible.
- Jet breaks with a conchoidal fracture (Fig. 2.2).
- Jet is totally rigid, so chain links have to be joined by cutting rather than by bending. Thus, in a jet chain, every second link will have two cuts, which are then glued together (Fig. 2.8).
- Jet jewellery has pins and clasps glued – not screwed – into place (Fig. 2.11).
- A greyish, polished sheen, or a brittle texture, suggests another type of coal.
- A very black piece that does not display the high lustre expected of jet may be jet of an inferior quality, for example from China (Fig. 2.10).
- A greyish, slightly rough, matt surface that is carved suggests bog oak (Fig. 2.5).
- Subjects that are Irish in theme such as a harp or a shamrock also indicate bog oak (Fig. 2.5).
- A material that is losing its colour cannot be jet or coal, and is likely to be vulcanite, which turns from greenish-brown to khaki with age (Fig. 2.3).
- Vulcanite is light and warm to the touch, as is jet.
- Under magnification vulcanite will not show complete evenness of colour and may display tiny spots. The slightly dimpled surface will also be visible (Fig. 2.6).
- It is unlikely that vulcanite will show fractures, but it may display worn edges.
- Lack of signs of carving suggest a moulded substance, as do air bubbles, joins, mould marks, rounded edges or unevenness of colour.
- Due to shrinkage in drying, moulded items may have slightly concave backs or facets.

Figure 2.10 Polished Chinese jet bead and Whitby jet brooch.

- Large items are very unlikely to be of jet but may be made of bois durci, which is also heavier than jet or plastics. Under magnification the wood flour content of bois durci may be visible.
- An object with clasps or pins screwed into place indicates vulcanite (Fig. 2.11).
- Chain links with only one cut in them cannot be jet as they would have to be slightly flexible to be opened and linked together. Jet is rigid, so flexible links indicate vulcanite or horn (Figs 2.7 and 2.9).

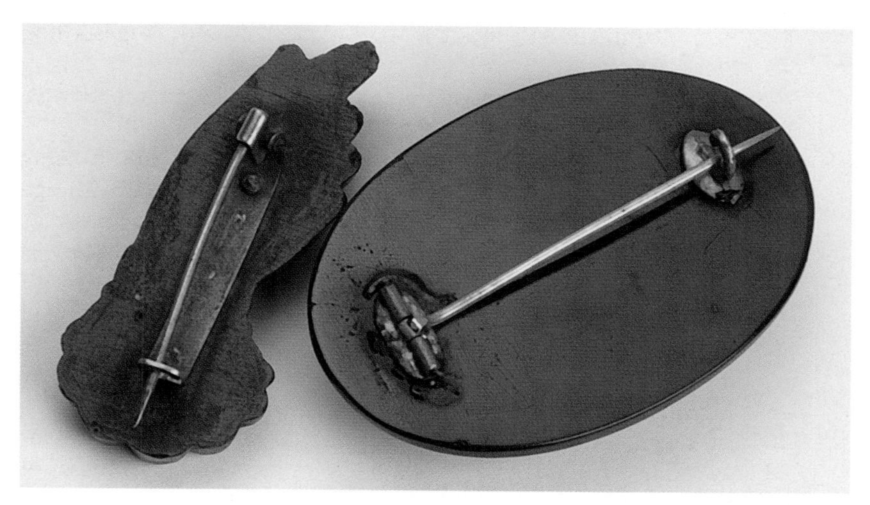

Figure 2.11 Brooch fittings: left, vulcanite with metal pins; right, jet with glue.

- Jet, coal, and vulcanite are opaque. An item that shows transparency in thin areas by transmitted light indicates glass or horn, or possibly celluloid or Bakelite.
- A moulded object with a high, vitreous lustre, which is cold to the touch and heavy is most likely glass (Fig. 2.4).
- A thin area of glass may appear transparent and reddish in colour by transmitted light. Cracks in the material will also show reddish glints of colour.
- Horn is very light and warm to the touch but may show signs of being moulded.
- On close inspection, possibly with magnification, it is usually possible to see the striations typical of horn. It may also be possible to see that it has been dyed (Fig. 2.7).

Tests

The streak test is destructive as it damages the surface of the material. However, it is usually possible to find an area where the necessary minute scratch will not show. The test is carried out by rubbing the item to be tested across a rough surface of neutral colour, for example unglazed porcelain or pale grey emery paper. An alternative method for the streak test is to take a minute scraping from the item with a very sharp blade, and to examine the resulting powder under a bright light and magnification. This is a more suitable method for testing beads, as the scraping can be taken from the drill hole. The colour of the streak or powder gives an indication of the nature of the material. Burning tests are also destructive as they can damage the surface of an item being tested. As with the streak test, the area to be burnt or scraped should be chosen with care.

- **Streak.** Jet gives a mid-brown streak. Coals produce a black streak – some material sold as jet from places such as Mexico and China also produce a black streak indicating that they are not true jet but another form of coal. Similarly, some materials produce a very dark brown streak, which is an indication of inferior quality jet. Bog oak gives an almost black streak. Vulcanite gives a brown streak. All other plastics such as Bakelite or celluloid will give a black streak, if any. Glass does not give a streak.

Note: When beachcombing, many items on the beach give an apparently brown streak when wet and should be retested when dry.

- **Burning.*** If there is still doubt about an item, it may be possible to take a very tiny scraping from an inconspicuous part and burn it. Jet and other coals will burn with a coal-like smell. Vulcanite will

Figure 2.12 Jet simulant bead, showing excessive damage from hot-point test.

give off a smell of sulphur – a smell similar to that of rotting eggs. Other plastics will give off an acrid smell.

- **Hot point test.*** A burning test can also be carried out by pressing a hot needle point against the specimen. It will only burn the surface of jet and coals, but will melt the surface of plastics. However, there is a danger of combustion if the plastic is celluloid, and also a danger of melting a large hole in the material (Fig. 2.12).

*Note: Care must be taken with the burning and hot-point tests as one of the early plastics (celluloid) *will ignite* when heated. It is marginally less dangerous to heat a scraping of the material to be tested than to stick a hot needle into the whole item.

A further test is possible to determine if an object is made from vulcanite:

- Placed on a piece of silver for a day or two, the sulphur content of vulcanite should leave a mark on the silver. The silver must be abraded clean as in a dishwasher, and not polished clean with a cleaner that leaves a protective film. Only vulcanised jet simulants will do this.

Conservation status and availability

Whitby jet is today collected only by beachcombing. It washes ashore after storms or appears on the beach after cliff falls have dislodged a seam. It is mostly collected by the few carvers living in and around the town, or by local youngsters who sell it to the carvers (Fig. 2.13).

Although there are some deposits a little inland, they are on private property, which is also a conservation area, and mining is not permitted.

Similar rules apply to Spanish jet, which is reported to be even more rare nowadays than Whitby jet.

It is believed that jet from areas such as the Ukraine, New Mexico and China are mined, but as already stated, these materials are of an inferior quality and often not workable.

Figure 2.13 Shale cliffs near Whitby.

PAST AND PRESENT USES

Jet has been used for millennia. In Switzerland and Belgium pieces of jet have been found that were apparently fashioned by flint stones. Its first use as jewellery is known to date back to the Bronze Age.

It was popular with the Romans who used pieces of highly polished jet as hand mirrors. They also fashioned it into hairpins, bracelets, rings and medallions, some of which were undoubtedly worn as amulets. It is thought that the Romans obtained some of their jet from what is now Turkey, though the better quality material came from England.

The Vikings who settled in Britain took a liking to jet, carving it into beads and animal figures. Some pieces were further adorned with a yellow pigment made with arsenic. The Vikings believed in jet's talismanic powers. This was probably due in part to the power of the arsenic.

Throughout the Middle Ages, jet was used mostly in an ecclesiastical context in Europe – for example, as rosaries and crucifixes – and it is thought that monks worked the local jet themselves. Spanish jet was worked mostly around Santiago de Compostela in northern Spain and the items made were sold to pilgrims as souvenirs.

The intense black colour, high lustre and warm feel of jet were undoubtedly what made it such a sought-after material. Yet its undiminished attraction in the Middle Ages may well have been partly due to a continuing belief in its special powers. It was thought that it could drive out devils and dissolve spells and enchantments. When heated it could ward off serpents. Furthermore, if a potion of powdered jet and water was given to a lady and acted as a diuretic, it was taken as proof that she was not a virgin.

In the sixteenth and seventeenth centuries jet was still being worked in France and Spain. By this time it was being combined with ivory and amber, and carved into religious artefacts and talismanic items, the latter sometimes with the addition of a Christian cross on the back. It was also still being worked in Whitby, but it was not for 200 years that a real industry evolved there.

In the early nineteenth century, in Whitby, jet was being turned on lathes, and cut, faceted and polished on large wheels. A fine, cream coloured mud from Derbyshire was used for the initial polishing on wheels, which were covered with walrus or porpoise hide, and the piece was given its high lustre with wool and iron oxide powder, before being finished off with a chamois leather. The workshops were full of highly inflammable dust and lit by kerosene lamps, with the result that several burnt down.

Collecting the jet was even more dangerous. It was done by hanging from the tops of the shale cliffs on ropes, and tunnelling in to extract the jet. The shale was very unstable and rockslides were commonplace.

Figure 2.14 Whitby jet necklace, late nineteenth century. *Jewellery by kind permission of Allison Massey.*

But jet was gaining in popularity and by the middle of the century there were around 50 workshops in Whitby, employing some 1500 workmen, plus a couple of hundred miners.

This was the time of strict observance of mourning in England, during which period only black could be worn. Furthermore, for deepest mourning, it had to be matt black, so the most suitable cloth was crêpe. This could be adorned with jet jewellery that had only been polished to a dull sheen. After a certain period of time, shinier cloths could be used and likewise jet jewellery that had been given a high polish. Jet is light and copious amounts could be worn without breaking any taboos. As women had to go into mourning for every relative, and as infant mortality was fairly high, many women found themselves more or less permanently in mourning. Also, when Prince Albert died, Queen Victoria went into mourning for 40 years, encouraging the fashion for wearing black. It gave the jet industry a tremendous boost.

Jet was at the height of its popularity. A display of Whitby jet at the Great Exhibition of 1851 in London attracted international attention and orders even came in from foreign royalty.

At the beginning of the twentieth century, early plastics began to come onto the market. They were much cheaper than jet as they could be mass produced and did not have to be hand carved and polished. Also, with deep mourning becoming a thing of the past, the demand for real jet waned.

Sadly, the very popularity of jet was also its downfall, as it had become irrevocably linked to mourning and was regarded as a very sombre form of adornment with sad connotations. The industry did not reinvent itself but continued with the same styles and designs, with the result that, even today, jet is regarded with little favour in spite of being a very beautiful material. Nowadays the few remaining carvers produce a few special and often beautiful items to order, but most of their work is for visiting tourists. Most popular are small items of jewellery such as jet cabochons set in silver as ear-rings or pendants.

In America jet has never gained the popularity it has enjoyed in Europe. The material found in Utah and Colorado has mostly been used by the Native Americans. They made it into mosaic and inlay, or carved it into small figures, combining it with turquoise.

3 Ivory

All mammal teeth or tusks are 'ivory', but the word is understood to denote those which are large enough to be cut, carved or otherwise fashioned in some way.

Ivory is most commonly thought of as coming from elephants, but the teeth or tusks from walruses, hippopotamuses, marine whales and dolphins, and members of the swine family such as wild boar and warthogs have also been used.

Structure and properties

All mammal species have different shaped teeth. Some of them have teeth that protrude from the mouth and cannot be hidden when the mouth is closed. These are called tusks and are usually used for foraging, fighting or manoeuvring, while teeth are used for tearing or masticating.

The structure of the tusks varies slightly with each species, though they all follow the same basic description.

- Put simply, a tooth or tusk is made of *dentine*, consisting mostly of calcium phosphate, with some organic material in the form of collagenous proteins. This makes up the bulk of the tooth or tusk.
- Inside this is the *pulp cavity* which caries blood vessels and nerves and which varies greatly in size among the different animals, some having much more solid tusks than others. Elephant tusks, for example, are solid for most of their length, while in narwhal tusks the pulp cavity occupies much of the length of the tusk.
- In some mammals the dentine is covered in a moderately calcified tissue called *cementum*. This can also be partially worn away. Cementum is the material that helps to hold the tooth or tusk in place in the jawbone.

- The upper part or crown of a tooth is covered with *enamel*, which is the hardest of all animal tissues and contains very little organic material. Hardness notwithstanding, this enamel is often worn completely away on at least part of the tooth and is seldom seen at all except on hippo teeth and boar tusks.

Dentine, enamel and cementum all have the same, or nearly the same, colour – typically an opaque cream – and they have the same lustre and can take the same, high polish. If there is any enamel on a piece of raw ivory it is usually removed before working.

Raw ivory may further display what is commonly referred to as bark. This is simply a discolouration of the outer layer, particularly noticeable in fossilised ivories, and is removed before working.

THE SPECIES

Elephants

There are two species of elephant: the **Asian** (also called Indian) elephant, *Elephas maximus*, and the **African** elephant, *Loxodonta africana*, which has two subspecies: the savannah (Fig. 3.1) and the forest elephant. All belong to the family Elephantidae.

Elephants are sociable and loving animals that live in matriarchal groups of mothers, their calves, and their grown daughters with their own offspring. All are very protective of one another, and especially of the young in the group. When the family gets too numerous, it splits into smaller groups, but they remain close by. The males either live alone or form their own bachelor herds.

The Asian elephant is generally the smaller animal of the two species, though there have been a few very large examples recorded. It has a slightly different body shape and much smaller ears in proportion to its body. Its tusks are narrower than those of the African elephant, generally carried only by the males and weighing a maximum of 40 kilograms. Today there are few Asian elephants found outside India.

The African savannah elephant is the largest extant land animal, standing at around three metres tall and weighing 5000–6000 kilograms. It has larger tusks than its Asian cousin. These can reach a length of 3.5 metres and a weight of 130 kilograms, though this seldom happens today. The female is a little smaller, as are her tusks.

The African forest elephant is a slightly smaller species, though still larger than the Asian elephant. Its tusks tend to grow straighter and parallel in a downwards direction.

Ivory from the African animals is a little softer than that of the Asian and is therefore easier to carve.

Figure 3.1 Young African elephant. Tanzania.

Elephant *tusks* are upper incisors embedded deep in sockets in the upper jaw. From the age of about two years when the deciduous tusks (milk tusks) fall out, the mature tusks grow throughout the life of the animal. They grow in a gentle, downward then upward curve and are solid almost throughout their length. There is pulp cavity only in the section of tusk that is within the jawbone. There is little if any enamel covering, and this is confined to the tip. The tusk is almost cylindrical and is covered in a narrow layer of plain dentine. Inside this, in cross-section, the main bulk of the dentine shows the typical 'engine turning' or cross-hatching pattern that is unique to elephant or mammoth ivories (Figs 3.2 and 3.12). This is a series of intersecting arcs caused by the internal structure of the dentine. In longitudinal section the arcs appear as uneven, slightly wavy lines. This wavy pattern of lines can appear on other ivories and is not confined to elephants and mammoths. African ivory is a pale, creamy colour, while Asian is a

Figure 3.2 Pepper mill, showing 'engine turning' pattern. Elephant ivory.

little whiter and more dense, and does not show the engine turning pattern so clearly.

As herbivores, elephants' other *teeth* are all molars. There are two teeth in wear on either side of the mouth, in both top and bottom jaws, and like the tusks, they also grow throughout the animal's life. They are worn away with chewing and are replaced by new teeth growing from the back of the mouth. This process is continuous so, if the elephant lives long enough, the full complement of six sets of teeth will have been worn away, and finally the animal will die of starvation. The molars are seldom used for ivory as they are not homogeneous. They are made up of folds of dentine and enamel and are stripy cream and white in colour (Fig. 3.13).

Elephants use their weight to knock down trees, and then use their tusks for stripping off the bark and young shoots. A fully grown African elephant may fell three trees a day in order to get enough to eat.

USES SPECIFIC TO ELEPHANT IVORY

Elephant ivory has always been the most popular of the ivories as it was the best and most versatile with which to work. Its bulk also meant that it was suitable for many things for which other ivories could not be used. For instance, at one time the billiard ball industry used vast quantities of elephant ivory as it was the only ivory large enough and homogeneous enough for the purpose. Elephant ivory has been popular for carving figures, and for large, hollow objects such as tankards.

Mammoths

Mammoths, *Mammuthus primigenus*, roamed huge areas of the globe for four million years. The last mammoths died out at the end of the last Ice Age, about seven thousand years ago. Contrary to popular belief, elephants are not descended from mammoths but are from the same species as mammoths, and so are cousins. The first to arrive in Europe probably did so about three and a half million years ago.

Mammoth fossils are occasionally discovered in various parts of the world, from South Africa to Siberia. In America fossils of the mastodon, *Mammut americanum*, a relative of the mammoth, are also found. They died out about ten thousand years ago. Mastodon tusks have sometimes been carved, but the tusks most commonly used today come from the mammoth. Fair quantities are unearthed in Siberia, during the two months of the year when the permafrost softens sufficiently to permit digging. Finds are also made in Alaska. The carcasses are deep frozen, which means that the mammoth tusks are fresh and well preserved – not mineralised – though they are usually stained dark brown on the outside. Much of the material found is too degraded to use, and what is used can have a tendency to crack. For this reason the tusks are nowadays used for small carvings or in mounted slices as jewellery, with cracks disguised by ornamental filler.

Mammoth tusks are not covered by trade bans, as the animal is already extinct and therefore not endangered. This makes mammoth ivory a popular substitute for elephant ivory today.

Early humans were fascinated by the beasts, which are the subject of numerous cave paintings and some very early carvings. They lived alongside the animals and hunted them, eating their meat, using their hide and putting their bones and tusks to use as building materials, or fashioning them into ornaments or weapons. When modern humans rediscovered mammoths melting in the Siberian summer thaw about four hundred years ago they were terrified, believing them to be underground-dwelling monsters that would spread death and destruction if they came to the surface. Their fear lasted until they discovered that the ivory tusks were valuable.

The tusks on an adult mammoth curved out and downwards, then in and upwards, gently spiralling in opposite directions and tapering towards the end. They were typically almost 3 metres long, though they could grow much larger. The biggest pair ever recorded was almost 5 metres long. Females also carried tusks, though these were smaller, less curved and less tapering.

In cross-section the mammoth tusk exhibits the same engine turning pattern as the elephant tusk (Fig. 3.12), and it can be impossible to tell them apart. Mammoth ivory can be a fraction darker in colour than elephant ivory, and does not take quite such a good polish. Also, the angles of intersection of the arcs in the dentine pattern tend to be narrower, at less than 90 degrees, as opposed to over 115 degrees in elephant ivory, though this can vary.

Odontolite

This is an extremely rare fossilised mammoth ivory that originated in southern France. For many years it was thought to have been stained blue naturally by the presence of the mineral vivianite in the surrounding soil. However, it is now believed that the ivory was deliberately heat-treated in the Middle Ages, to resemble the mineral turquoise. It is easily mistaken for turquoise, but close inspection reveals the typical dentine patterning of mammoth tusk. Today odontolite is only seen in museums or private collections.

Walrus

The walrus, *Odobenus rosmarus*, is a sociable arctic animal that lives in large colonies. The North Pacific animals are larger than those of the North Atlantic, and their herds are flourishing. The North Atlantic herds are dwindling and coming under threat, as a result of over-hunting.

The walrus belongs to the Pinniped family, which also includes seals and sealions. A mature Pacific male walrus can weigh 1200 kilograms, and measure 3 metres in length. Although ungainly on land, the walrus moves swiftly and elegantly in water where it spends over half its life.

Walrus *tusks* are modified upper canines, which grow downwards from the upper jaw in a slight curve. They are unlikely to be used for foraging on the sea bed, but rather are used as a cross between runners and stabilisers that protect the head as the animal trawls the sea bed in the dark, searching for food. Evidence of this is seen on mature tusks where the dentine is well worn on the front. They are also used as ice-picks with which the animal can haul itself up onto slippery ice flows, and for fighting among the males of the species. In order to lift such a weight the tusks are very solid and can reach up to 1 metre in length.

Figure 3.3 Inuit carving in walrus ivory, showing primary and secondary dentine. Mid-twentieth century.

The females also carry tusks though theirs are smaller and seldom measure more than 60 centimetres.

The tusk has a layer of cementum covering the dentine, and enamel at the tip, though both may be worn away. The cementum is also prone to cracking along the length of the tusk. In cross-section these cracks appear as radial lines around the edge.

Diagnostic for walrus ivory is that it is the only ivory to contain primary and secondary dentine. The former displays very finely packed, concentric lines and makes up the bulk of the tusk. The latter has a bubbly appearance like that of tapioca or rice pudding (Figs 3.3 and 3.16). It fills the centre of the tusk all the way through from the tip to a small pulp cavity in the part of the tusk attached to the jawbone. This secondary dentine looks more fragile than the primary dentine, but it has the same hardness, and the material can be carved and polished with little regard to the different textures (Fig. 3.3).

The rest of the walrus's teeth are small and peg-like, and only about 5 centimetres in length. They tend to have a proportionally much thicker layer of cementum and a very small area of secondary dentine inside the primary dentine. The secondary dentine may simply appear as an irregularity rather than a bubbly mass. Similar to other ivories, walrus ivory is creamy white throughout.

It is also found and sold in fossilised form and can retain its creamy colour, though it usually appears a pale brown shade, especially around the edges (Fig. 3.16). It is found in the permafrost of Alaska and Siberia and is – like the mammoth ivory that is also found in those areas – frozen rather than mineralised. Fossilised walrus ivory takes an excellent polish, but is mostly carved as a curiosity.

Uses specific to walrus ivory

Walruses in Siberia are hunted for their meat, which is used to feed farmed foxes bred for their fur, but there is a ban on the sale of the ivory. The Inuit of North America and Greenland are allowed to hunt a quota of walruses as their cultural right. They eat the meat, use the skin as a strong and durable leather, and are permitted by licence to carve and sell the ivory.

Walrus ivory was much used in medieval times when elephant ivory was difficult to obtain. It is therefore common to find old ivories carved from walrus tusk in churches and in museum collections.

After about AD 1300 trade was resumed with East Africa and elephant ivory again became the preferred material as it is slightly softer to carve, the tusks are larger, and the structure of the dentine is uniform throughout.

Cetaceans

Cetaceans include whales, dolphins and porpoises.

Extant cetaceans belong to one of two suborders. The dolphins, porpoises and the toothed whales belong to the larger suborder Odontoceti. The baleen whales belong to the smaller suborder Mysticeti. In place of teeth the mysticetes have hundreds of overlapping keratinous plates called baleen, which hang from their upper jaws, where they form a sort of brush which acts as a sieve (See 'Baleen' in Chapter 12, 'Miscellaneous organics').

Sperm whale and orca

It is most usual to come across ivory from the sperm whale (cachalot), and the orca (killer whale). These two are here grouped together due to the similarity of the structure of their teeth.

The **sperm whale**, *Physeter macrocephalus*, is the largest of all toothed whales and the largest extant carnivore. It lives in deep waters in most of the world's oceans, though it is only the mature males that visit Arctic waters. It is recognised by its huge, square head.

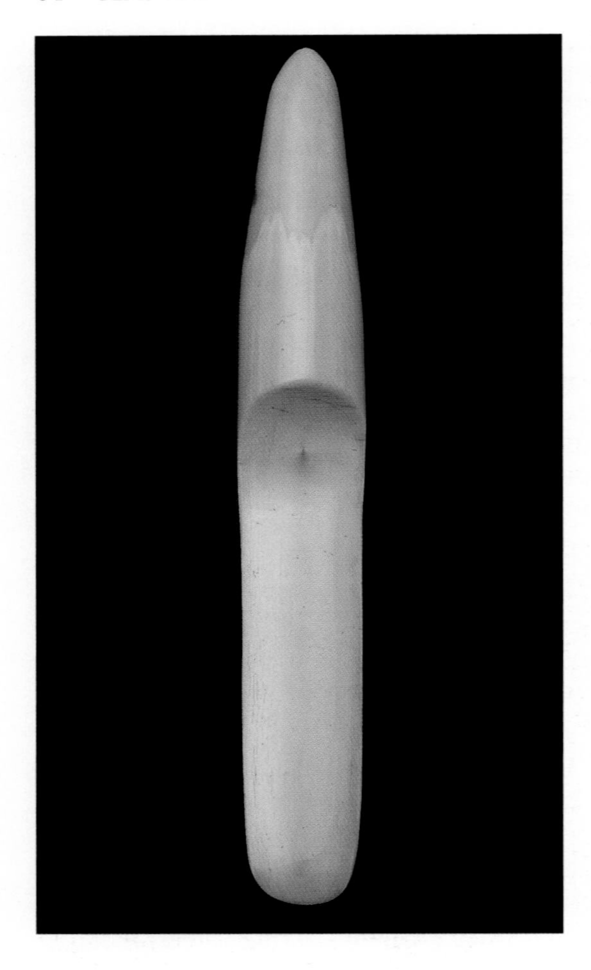

Figure 3.4 Letter opener. Sperm whale ivory.

A male can grow to 16 metres and weigh up to 44 tonnes when fully mature, while a female is usually less than 11 metres long, and much lighter. The new-born calves weigh 1 tonne. Females reach maturity at 45 years, and only calve every five years.

Sperm whales have 20–25 peg-shaped *teeth* on each side of the lower jaw only, those of the upper jaw remaining undeveloped. The teeth emerge late, and in some females never emerge at all. They can grow to around 25 centimetres in length and weigh more than 1 kilogram.

The **orca**, *Orcinus orca*, though commonly called the killer whale, is not in fact a whale but is the largest of the dolphins. It is easily recognised by its very distinctive black and white markings. It prefers cold seas, especially the polar regions, but it will venture further afield. It is much smaller than the sperm whale but is greatly feared by other cetaceans and sea animals as it is a versatile predator, attacking anything from fish and birds to the largest whales. Orcas catch this large prey by working together and hunting in groups. Although they may tease their prey before killing it, orcas kill to feed, rather than for pleasure.

An orca has 40–50 teeth: 10–13 on each side of the top and the bottom jaws. The teeth interlock neatly when the jaw is closed. Orca teeth are slightly smaller and narrower than those of the sperm whale and tend to curve slightly inwards and backwards to facilitate feeding off larger prey.

The teeth of both species are solid for most of their length and are enamel tipped. The rest of the tooth is covered with a layer of cementum. In cross-section they appear round or slightly oval, with a transition ring evident between the cementum and the dentine. The dentine displays fine, concentric rings, with a dark spot in the centre (Fig. 3.4). In longitudinal section the cementum is still clearly visible on the outer

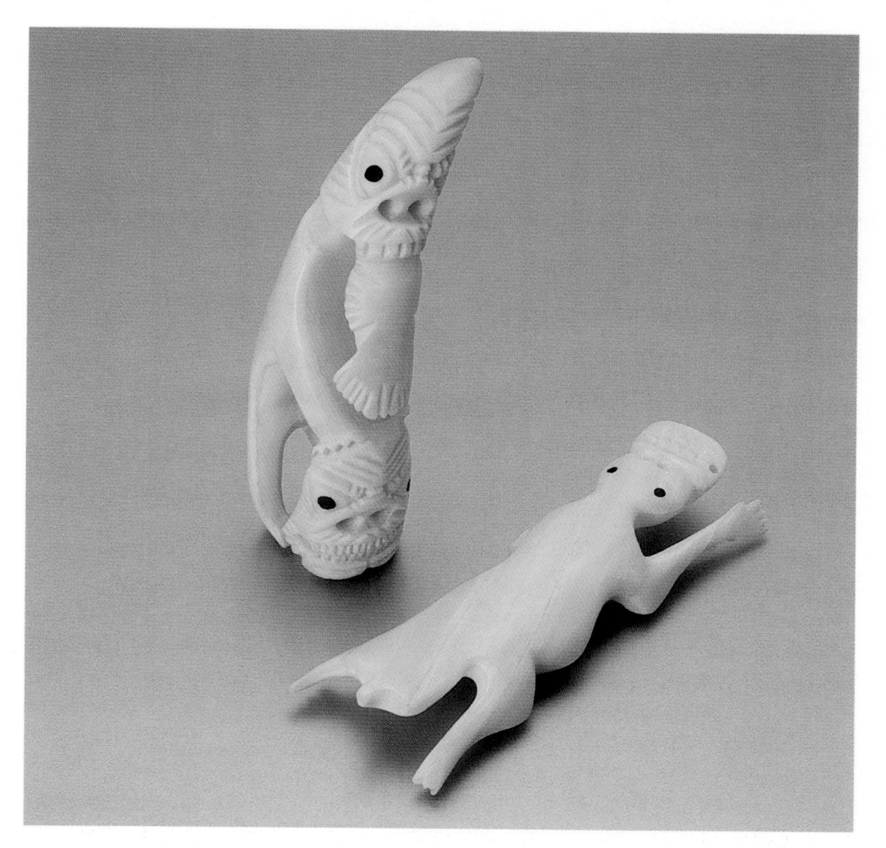

Figure 3.5 Two tupilaks from Greenland. Mid-twentieth century. Spermwhale ivory (standing) and narwhal ivory.

edges, while the dentine displays conical striations following the outline of the tooth, and the dark spot shows as a dark line through the centre of the tooth (Figs 3.5 and 3.4). The ivory is a creamy colour, though when fresh the cementum appears paler.

Narwhal

The narwhal, *Monodon monoceros*, is an unusual toothed whale known for its single, long, spiralled tusk. It lives in the North Polar regions, seldom venturing further south than the Arctic Circle. It is not a large whale, reaching only a maximum of 5 metres in length at maturity.

Narwhals have two teeth in the upper jaw. In the male the left tooth develops into a protruding *tusk* when the animal is aged about one year. This grows straight forwards, in line with the body axis, to around 2 metres in length, spiralling anti-clockwise along its length (Fig. 3.14). Very occasionally both teeth develop into tusks but the left tusk is always longer. Very rarely will a female produce a tusk, and when she does it seldom grows to much more than 1 metre in length. The teeth that do not grow into tusks remain virtually undeveloped and do not erupt through the gums.

The purpose of the tusk is baffling. It is unlikely to be used to break the arctic ice as the whale does this by head-butting. The tusk may be used for trawling the seabed to raise food, but its main purpose is thought to be as a weapon for jousting with other males, and as a status symbol in the group.

The tusk has a pulp cavity running along most of its length and is therefore hollow and rather brittle. The cementum coating often shows some cracking, which follows the spiral of the tusk. In cross-section the tusk is round but with peripheral indentations following the spiral shape. There is a clear transition ring between the cementum and the dentine, and the dentine displays fine, concentric lines. A cut piece of the ivory is convex on the outside, and displays diagonal lines on the cementum following the spiral of the tusk (Fig. 3.5). The inside often appears concave as the layer of ivory is thin and the pulp cavity is large.

Uses specific to cretacean ivory

It is said that the teeth need to be taken from the whale quickly after it has been killed, and that the ivory is best carved fresh. Teeth found on an animal that has died naturally can turn greyish and become brittle.

Cetacean ivories are mainly associated with the whaling industry and scrimshaw (the whalers' art of carving available materials, especially ivory). The teeth were also valuable to the Inuit, as they could be made into utensils, and carved into such items as toys and talismans, as well as harpoons, or harpoon points. Narwhal ivory has limited use as a material for carved or decorative purposes because of its unusual shape and because it is not solid, though it has been used to adorn items such as tankards or casks, where lengthways slices of tusk are applied. The undeveloped narwhal teeth, which can measure up to 20 centimetres long and do not spiral, were useful for carving into such items as crochet hooks or tobacco pipes.

A whole narwhal tusk is very spectacular, and, as part of the appeal of narwhal tusks was the belief that they were unicorn horns, they were usually left intact. In this form they were used for bishops' croziers, and shorter lengths were used for canes or in furniture. A notable example is the throne which was made for the coronation of Christian V of Denmark in 1671.

Hippopotamus

The hippopotamus, *Hippopotamus amphibius*, affectionately referred to as the hippo, is in fact an exceedingly dangerous and aggressive animal, and is the cause of far more human fatalities in sub-Saharan Africa, where it lives, than either lions or elephants. It is distantly related to the swine families, and is the closest living terrestrial relative of the great whales.

Its name derives from the Greek hippopotamos, meaning river horse. At one time widespread, the hippo probably originated in Africa and spread to Europe and Asia, but never as far as America or Australia. It now lives only in Africa.

A very large hippo can weigh up to 3200 kilograms, and stand 165 centimetres at the shoulder. It is a nocturnal animal and is vegetarian, eating up 40 kilograms of grass per night, and spending most of the daytime lying in water to keep cool. It is a poor swimmer, so when viewed apparently floating in water the animal is probably standing or lying on the bottom. If it needs to cross the water it walks, even if this means walking under the water. It uses its teeth for feeding and for fighting. When it 'yawns' it is actually giving an aggressive warning signal, rather than expressing tiredness.

The hippo is born with deciduous teeth (milk teeth) but these fall out when it is a few months old. The mature teeth include two pairs of *tusk-like canines*, one pair in each jaw. The bottom pair are curved and point forwards, though they do not protrude from the closed mouth. In a large animal these can be up to 70 centimetres long, with 30 centimetres exposed and 40 centimetres of root, and can weigh up to 3 kilograms. In cross-section very finely packed lines may be observed in the dentine following the shape of the tooth which is slightly triangular, though the dentine is very dense and the lines can be difficult to discern. A small gap called the interstitial zone often occurs at the centre of the dentine as it converges during growth, but otherwise the teeth are solid almost to the base (Fig. 3.17).

The top canines are smaller and round or oval in cross-section, with an indentation on the inner surface. In cross-section the same fine lines may be observed, and again there is often a clear interstitial zone. The exterior of all the canines have longitudinal ridges. The canines are covered in enamel, but this is worn away from the back of the lower teeth, and from the front of the upper teeth, where the two pairs rub against each other. The remaining enamel is usually removed before carving.

Between the canines are four peg-like *incisors* in each jaw, the bottom teeth are triangular in cross-section and the top teeth are round, and shorter. In cross-section they also display fine, concentric lines in the dentine, and have a small dark dot at the centre. Both the canines and the incisors grow throughout the life of the animal.

Finally the hippo has three molars and three pre-molars in wear on either side of the upper and lower jaws, but these are not used much for carving.

Hippo ivory is the hardest of all ivories. It has very little organic content, is very dense, and is creamy white.

Uses specific to hippo ivory

Because it is extremely dense, hippo ivory is resistant to staining, and has therefore, in the past, been preferred to other ivories for the manufacture of false teeth. It was not much used before the nineteenth century, as until then the enamel was too hard to remove with the tools available.

Wild pigs or suids

Often forgotten as a source of ivory, there are various species of the pig family, Suidae, that bear tusks. These include wild boar, warthog, bush pig, babirusa and others. They are found worldwide with the exception of Australia, though they are no longer very common in North America.

The various species of wild pig vary in size and colour, some having fur and others being almost bald. They are all barrel shaped like our domestic pigs, and all have the typical snout.

They are gregarious animals, mostly living in groups. They love water and wallowing in mud, though they keep themselves very clean. They do not kill for their food, but live off carrion and forage for vegetation with their snouts, seldom using their tusks for this purpose. Contrary to popular opinion they are only dangerous when injured or threatened.

The *tusks* of the wild pig are used for fighting and can inflict very nasty wounds, being thin and pointed. There are two pairs – one tooth

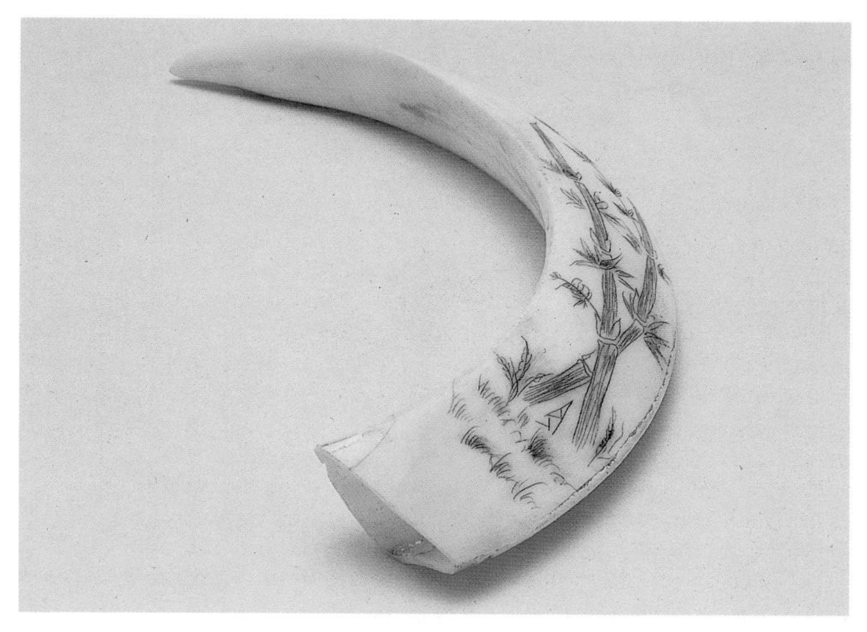

Figure 3.6 Boar's tusk with engraved and coloured pattern.

on each side of the top and bottom jaws – which are canines. The lower pair juts out at right angles from the jaw and curves slightly upwards, while the upper pair are bigger and curl out and back into large tusks.

The wild boar, *Sus scrofa,* was once native to the whole of Europe, but died out in some of the northern countries, including the UK. It is thought that their recolonisation of the UK followed the Great Storm of 1987 when wild boar escaped from rare breeds farms. Like all suids, they are prolific breeders and are now happily multiplying in the south eastern counties.

Although piglets are pretty little things, not everyone considers the adults to be attractive creatures, and the African warthog, *Phacochoerus aethiopicus,* is a prime example. The spectacular tusks of the warthog add to its unusual appearance.

The barbirusa of Indonesia carries the most elaborate of all suid tusks in that the top pair emerges through the skin at either side of the nose, rather than from the mouth, and then curls up and back into an almost complete circle.

Suid tusks are hollow for much of their length. They are also slender and all of them curve to some degree (Fig. 3.6). In cross-section they can be squared or triangular and show irregularly spaced concentric lines of dentine. Much of the exposed tusk may be covered with enamel and have furrows along its length. The creamy colour is not always even.

Figure 3.7 Model of the frigate *Norwegian Lion*, by Jacob Jensen Nordmand, 1654. Elephant ivory. *Rosenborg Castle, Copenhagen, Denmark.*

Uses specific to suid ivory

Due to the size and shape of suid ivory it is seldom carved. The tusks are usually left whole or may be polished or inscribed as a decorative item. They can be used as handles – for example, for knives – or threaded as necklaces or bracelets, to be worn as decoration or as trophies.

Figure 3.8 Detail of ivory inlaid cabinet, with etched pattern. English. Early twentieth century.

TREATMENT AND USES

- Depending on the natural shape of the ivory being used, it can be carved into various objects such as figures, pictures, netsuke and ecclesiastical items. It can be turned on a lathe to make items such as round boxes or billiard balls. Both carving and turning have often been used on the same item.
- Ivory is usually polished after working, giving a beautiful and lasting lustre.
- Elephant ivory can be cut in a very thin layer by rotating a piece of tusk against a blade, and slicing it finely in longitudinal section. This was used for inlay or piano keys.
- Thin strips of elephant ivory can be soaked in water and gently bent into a curved shape. Examples of this are the sails in carvings of sailing ships (Fig. 3.7). Ivory cannot be moulded.
- Ivory can be used as an inlay, sometimes with a pattern etched into it for further decoration (Fig. 3.8).
- Ivory can be inlaid with materials such as mother-of-pearl, coral, tortoiseshell or ebony. This has been popular in Japan, where it is called Shibayama work.
- With the exception of hippo ivory, all ivories can be stained. This is often seen in chess pieces or knife handles.

- Ivories can be etched and inscribed, and the resulting patterns stained with black or colours to make them more visible (Fig. 3.6). The most notable example is scrimshaw.
- Ivory can be painted, as, for example, when used for miniature portraits.
- In some instances ivory is stained to 'age' it, giving it a slightly yellow look. This treatment can make it extremely difficult to judge whether the piece is genuinely old, or new and possibly illegal. The methods of attaining this effect have been used for centuries, and included burying the ivory in damp hay or dung, or soaking it in tobacco juice or tea.
- Ivories can be bleached, if they have darkened, by dipping them in turpentine and leaving them in strong sunlight. This dries out the material and it can crack.
- Ivory darkens when not exposed to light.
- When handled constantly, ivory gains a beautiful polish. However, the acid in human perspiration darkens the material, as do the natural oils (which are present on human skin) when they oxidise. The material can turn dark brown as a result.
- The material can be artificially cracked to 'age' it. It is worth noting that natural cracks only occur following the grain of the ivory, and appear *after* carving or other decoration.
- Occasionally an item has been given a coating of some type of plastic. Cosmetic's containers, for example, could be stained by their contents and a coating of plastic prevents this.
- Ivory dust resulting from working the material can be burnt to produce the blackest of all black pigments called 'ivory black'.
- Ivory is a poor conductor of heat, and therefore excellent for use as handles on silver tea or coffee pots.

SIMULANTS

- **Bone** is the most common simulant of ivory. In small items or as inlay it can be difficult to tell which material has been used as both bone and ivory appear much the same colour and have many similar properties. However, bone contains none of the structural patterns of ivories, for example the engine turned pattern of elephant and mammoth ivory or the tapioca pattern of the secondary dentine in walrus ivory. Instead it has the black dots or lines of the Haversian canals (nutrient bearing canals) (Figs 4.2 and 4.3).
- **Plastic.** Modern plastic simulants are sometimes called 'faux ivory'. Plastic has been used to simulate ivory for almost 100 years, indeed the early plastics such as casein and cellulose nitrate were some of

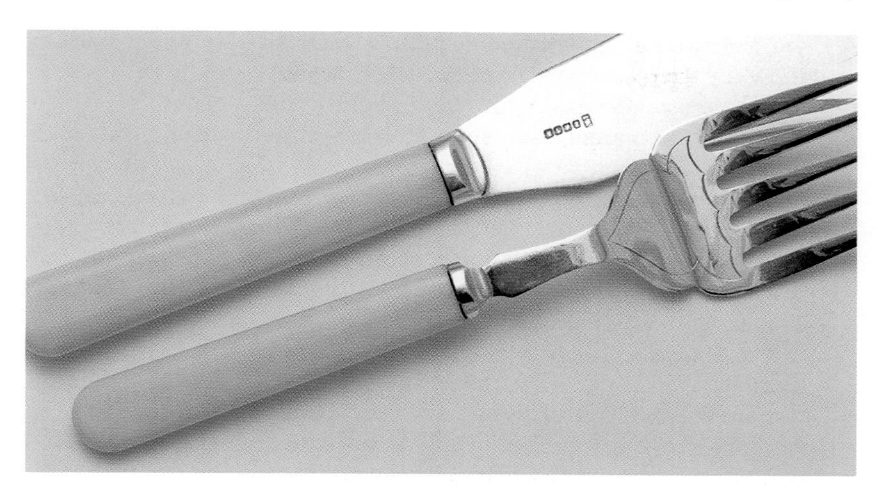

Figure 3.9 'Ivorine' plastic handles, imitating elephant ivory.

Figure 3.10 Plastic imitation of sperm whale tooth with scrimshaw.

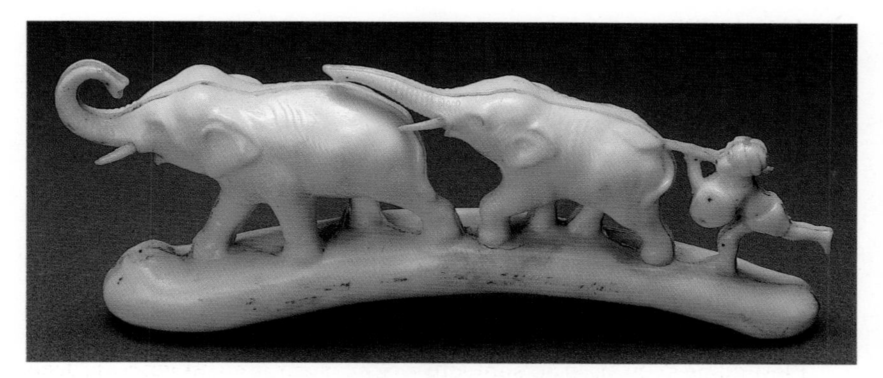

Figure 3.11 Moulded plastic imitating ivory carving.

the best imitations. Although plastics have been produced which show a pattern similar to that of elephant ivory (made by alternating narrow blocks of opaque and slightly translucent plastic), the pattern is too regular and square in cross-section, and shows almost parallel lines when viewed lengthways (Fig. 3.9). It was often used for the handles of such items as cutlery, carving knives and fish servers, and sold under names like 'Ivorine'.

Imitation scrimshaw is made of plastic – for example, epoxy – which is moulded in a silicone rubber mould that is flexible. This enables it to be pealed off the moulded object, and eliminates the risk of marks from joins in the mould. The plastic is weighted to simulate ivory (Figs 3.10 and 3.11).

- **Reconstituted material** is made by combining the dust from working ivory or bone with a filler and moulding it. This results in the tell-tale signs of moulding: possible bubbles, joins, and so forth, and none of the signs of having been carved. It has no distinguishing features and has either a homogeneous colour throughout, or possibly swirls of colour. It has little, if any, fluorescence.

  Re-constituted material made with powdered ivory is seldom seen, especially in modern pieces, due to the ivory trade bans. It is commonly made with bone.
- **Vegetable ivory.** Various large nuts, especially tagua nut, have long been used as a simulant for ivory. With the present bans on animal ivory, tagua nut is again popular. It consists of cellulose and can be carved, turned and dyed. Viewed under magnification it bears no resemblance to ivory, but has a densely packed, spotted appearance (see Chapter 12, 'Miscellaneous organics') (Fig. 12.8).
- **Minerals** of a creamy white hue, for example 'ivorite', could be confused with ivory, especially when carved. Examination would reveal that they are cold and hard to the touch, heavy, and have a complete lack of structure.

Tests and identification

Chemical tests are not used on ivory, as they would damage it. Identification relies on the appearance and feel of an object.

Visual examination

- Ivory is best examined by the naked eye in strong light. A 10× lens may also be useful.
- Ivory has a typical, warm, creamy colour, a good lustre and a warm, silky feel – there is nothing hard and brittle about the feel of ivory.

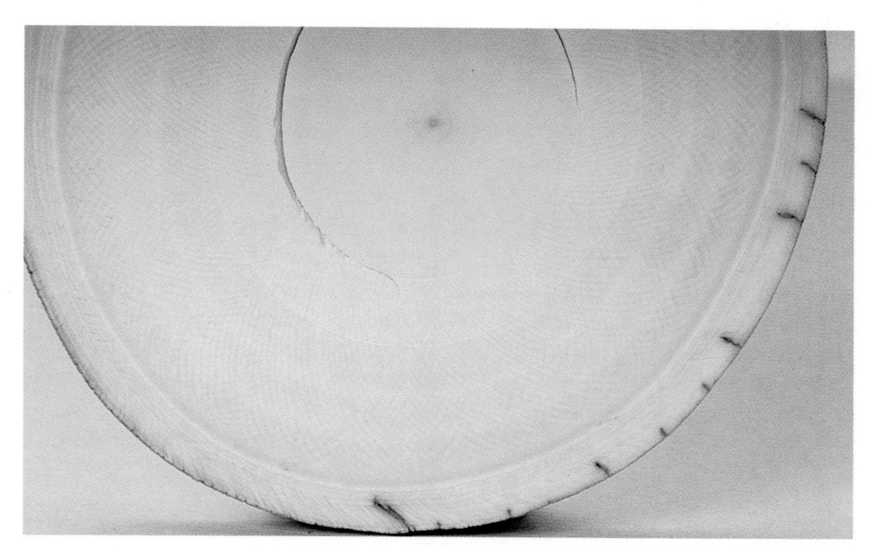

Figure 3.12 Cross-section of mammoth tusk, showing 'engine turning' pattern.

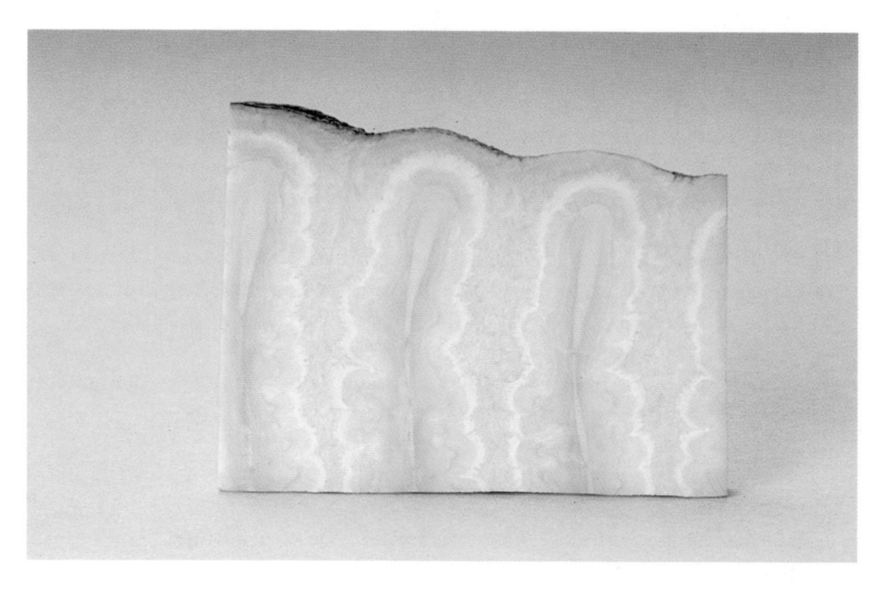

Figure 3.13 Cross-section of elephant molar, showing folds of dentine and cementum.

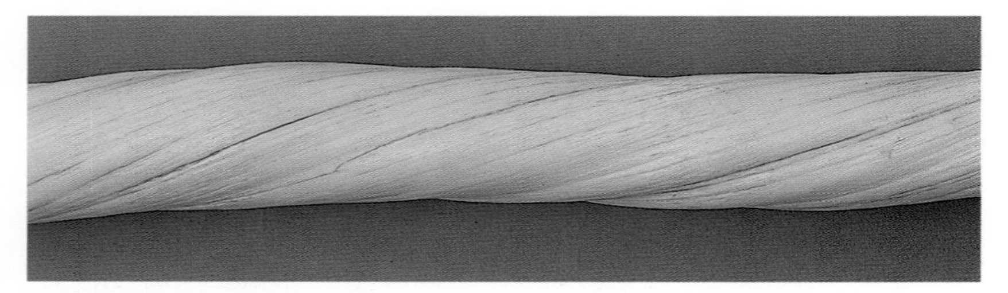

Figure 3.14 Section of narwhal tusk showing spiral.

Figure 3.15 Cross-section of narwhal tusk.

- The diagnostic for certain ivories is their structural patterns, for example the 'engine turning' pattern in elephant or mammoth ivory, the tapioca pudding look of the secondary dentine in walrus ivory, the diagonal pattern on narwhal ivory and the concentric lines and dark central spot of sperm whale ivory (Figs 3.4 and 3.12–3.17).
- In larger carvings or objects, the curvature of a tusk may be detectable. Examination of the structural pattern is needed to determine which type of tusk has been used.
- Bone does not take as good a polish as ivory.
- Small black dots in cross-section, appearing as lines in longitudinal section, are diagnostic for bone (Figs 4.2 and 4.3).
- Plastics have a different feel with a less velvety surface.
- The surface of plastics scratches easily and dulls with age, whereas, although ivory can be scratched, it retains its high polish.
- Even, fine, parallel layers of opaque and slightly translucent mater-

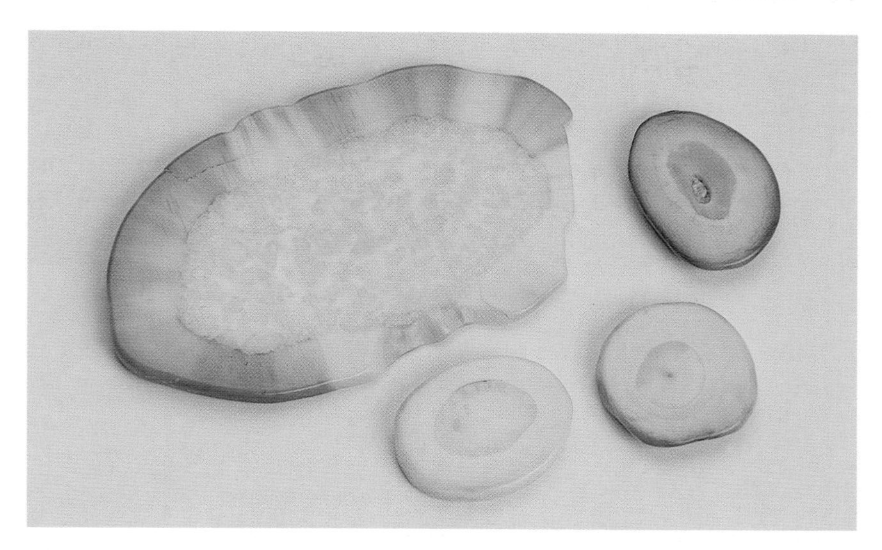

Figure 3.16 Fossilised walrus ivory, cross-sections.

Figure 3.17 Cross-section of hippo tusk, showing interstitial zone.

ial, seen in longitudinal section (or seen as blocks in cross-section), indicate a plastic imitation of elephant ivory (Fig. 3.9).

- A completely uniform colour, swirls of colour, or complete lack of structure indicates a plastic or reconstituted bone or reconstituted ivory. This is difficult to see in very small items.
- The occurrence of air bubbles or mould marks is proof of plastic or reconstituted ivory or bone (Fig. 3.11).

- Unless specially weighted to resemble ivory, plastic is a lighter material.
- Plastic may be flexible to some degree, while ivory is rigid.
- The intricacy of the work may be some indication of its authenticity. The better the workmanship, the more likely it is that the object is made of ivory. Bone has sometimes been intricately carved, but never plastic.

Tests

- **Ultraviolet light.** Ivory fluoresces chalky blue in UV light. Bone fluoresces in much the same way as ivory in UV light, as does the early plastic, casein, but other plastics are inert.

Conservation status and availability

Nowhere has the subject of conservation been more emotive than in connection with the elephant. It has also, in some ways, been a success story, as elephant numbers have increased in many areas where the wildlife has been confined for its own protection. Ironically, this has resulted in overpopulation in some small areas, which means the habitat of the black rhino – itself an animal on the brink of extinction – is being threatened in those areas by the elephants' huge appetite for trees.

Left to nature, the elephant population has, in the past, always been kept under control by illness, drought and other natural occurrences. It is a sobering thought that, thanks to the popularity of ivory, there is now only approximately 15 per cent left of the elephant population that once roamed in Africa. The hunting ban has been a success but with such low population numbers the elephant is by no means safe from extinction. For this reason the ban on the trade in fresh elephant ivory remains. In a few countries, ivory collected from dead animals, or that has been stockpiled during the complete ban on trade, has now been released for sale, but only under extremely strict controls.

Hippos are still hunted for their teeth, and for their skin, which is 25 millimetres thick. Though not at present under threat of extinction they may become so due to the popularity of their ivory, following the lack of elephant ivory. Nowadays hippo ivory is imported by Japan and China for carving. Elsewhere, for example Europe and America, it is covered by the total ban on the import of any ivory, so is not available.

As regards marine ivory, for whatever reason, there is concrete evidence that the Arctic ice cap is getting smaller and the surrounding waters warmer. This is having a detrimental effect on the local ecology and in turn threatens the wildlife in the region, including the animals that give us marine ivory.

Most Arctic ivory is covered by a ban on hunting, yet the Inuit in Alaska and Greenland are allowed to catch a quota of walrus – which is at present listed on CITES Appendix II – each year as part of their cultural rights. The raw ivory can be used, worked and sold only by local people, and only under strict control. This rule makes it possible to buy walrus ivory items legally in America. It is also possible to buy walrus ivory in Denmark (as Greenland is a protectorate of that country), where it has been imported with full CITES documentation. Exporting the material from these countries, however, may not be permitted, and would require further documentation.

Past and present uses

Ivory has been regarded as valuable and desirable for at least 7000 years, and there have been finds of mammoth ivory implements dating back to the Stone Age. As ivory is more durable than many of the organics, we are able to trace its history more easily. It has never lost its popularity, but, as fashions changed, it has been used for different forms of adornment. The very best craftsmanship has always been lavished on this beautiful material, and it has enjoyed a status never attained by any of the other organic materials. For example, many thrones have been made of ivory.

The palaces and temples at Nimrud in Assyria (now northern Iraq), which were first excavated 100 years ago, were full of ivory carvings and reliefs used as panels on furniture. Many showed signs of having been inlaid with stones or covered in part with gold leaf. As the area had been sacked, much of the ivory that was excavated was found to have been badly damaged. The passing of time has also degraded it, so that it looks largely like old, carved wood, but it is incredible to see the detail of the carving and, on some pieces, there is still a hint of their original high polish. These ivories are now to be found at the British Museum in London, the Metropolitan Museum in New York and in Baghdad.

India had their own elephants, and their history of ivory carving goes back about 4000 years. The earliest uses appear to have been limited to small objects such as games pieces or jewellery, but there is evidence that India was exporting ivory as early as 500 BC. By the year AD 200 ivory was being used as decoration on items used by royalty, such as sword hilts or furniture. Plaques of ivory were carved to adorn furniture (especially palanquins, thrones, ornate cabinets and bedsteads), and the material was used as an inlay, or even as a veneer covering the

Figure 3.18 Card case. Japanese. Late nineteenth century. Elephant ivory.

whole item. Further, the ivory was sometimes engraved and the pattern stained black, for a more ornate effect. The wood used was often rosewood, and the carvings often depicted people and animals.

The art of inlaying gem materials with gold and more gem materials was perfected in India. Ivory was an obvious base for this work, and there are many examples of carved ivory figures or jewellery that have been inlaid with gemstones.

The earliest examples of carved ivory come from China, and are dated around 5000 BC. They are a few, simple pieces such as perforated plaques. Little exists from such early times, but we know that, a couple of thousand years later, ivory jewellery was put into burial sites. For many centuries it was used for jewellery and as inlay for furniture. Only later – around 700 years ago – did the art of carving it into figures and ornaments develop.

The true masters of the art of carving have always been the Chinese, and, with their reverence for ivory, the carvings they have produced in this medium are unsurpassed in quality. Beautiful statues, ornamental objects with carving so intricate that it resembles lace, the famous balls within balls made from a single piece of ivory, even chains made with solid links, all these and more have been produced over the years.

Japan was strongly influenced by China, and their craftsmanship is almost on a par with that of the Chinese. Originally, ivory was used as inlay, sometimes together with tortoiseshell, silver and gold. And, like the Chinese, the Japanese later developed the art of carving it.

Figure 3.19 Carved ivory panel depicting an eagle: the symbol of St John the Evangelist. Carolingian. Ninth century. *Victoria and Albert Museum, London.*

Netsuke – toggles worn on the clothing – are probably the best known Japanese carved items. Also made were figures, seals, and boxes. Later many of the items produced were made specifically for export to the west (Fig. 3.18). As in India, ivory in Japan has also been used as a background and inlaid with other precious materials such as coral and tortoiseshell. They called the style Shibayama, after the man who was a master of the craft.

The museums around Europe have plentiful examples of ivory carvings dating from early medieval times. Most of these are ecclesiastical carvings such as portable altars, and many are of walrus ivory, because the elephant ivory trade into Europe was interrupted for 400 years or so by the Islamic conquests. Again, some of the pieces are aged and cracked and look somewhat woody, with colours varying from pale beige to brown, but still they retain their intricacy of carving and the hint of a polished surface. Others are in excellent condition and even retain some of the paint with which it was customary to adorn the carving (Fig. 3.19).

About this time another form of ivory was being used in Europe – especially in England. Beaver teeth were worn as amulets. Unlike today, the animals were then abundant, but the belief that lay behind the wearing of the amulets is unclear.

In the seventeenth and eighteenth centuries the various European royal courts had their own master ivory carvers who produced magnificent figures, reliefs, ornaments and trophies. These were frequently mounted in heavy silver or silver gilt, and were added to the royal collections to be used as decoration for banquets, or as gifts, or whatever caught the whim of the regent of the time (Fig. 12.12). These master carvers often travelled, learning their skills in one country and moving to another for work. One such carver was Magnus Berg, who was born in Norway at the end of the seventeenth century, but finished his training in Italy before moving to Copenhagen, where he worked at the royal court. Many of his magnificent pieces can be seen at the Rosenborg Museum in Copenhagen.

Later ivory became much more readily available to the general public, and by the start of the twentieth century was immensely popular in Europe. It was used for such items as dressing table sets, writing sets, card cases, games and knife handles – all useful objects but with ivory giving them the added touch of luxury.

As already mentioned, the billiard ball industry used vast amounts of elephant ivory. Only three, good-quality balls could be made from a single tusk. As they had to be perfectly round, the production of the balls was very specialised and involved months of drying the material before and during manufacture. More large quantities went on piano keys, and on cane or umbrella handles. This fondness for ivory lasted until the elephant hunting bans of the 1980s.

In America ivory was never very highly prized. A little was in use, mostly in the form of handles on imported English silver tea and coffee pots. The luxury of ivory had largely been left behind in Europe by the people who went to America, and the Native Americans, with the exception of the Inuit, did not have access to ivory, so did not use the material at all. But America is renowned for one form of ivory use, and that is scrimshaw – the whalers' craft.

The art of scrimshaw was actually started by European whalers, but quickly spread across the Atlantic and is now generally considered an American art. It came about because the men on the whaling vessels had a lot of time and little to occupy themselves with in between sightings and catches of whales, so they used the by-products of the whaling industry to pass the time. Marine ivory, bone and baleen could all be carved or etched and in some way decorated to make useful and attractive objects.

Some items that were made were functional and used on board ship, but many were made as gifts for wives or girlfriends back home. Typical items were pastry cutters, bobbins, and sewing or knitting implements, made from whale teeth. Sometimes teeth were left whole, pictures were etched on the surfaces, and they were mounted as purely decorative objects (Fig. 3.10). The most common illustrations etched onto the teeth were sailing vessels, patriotic scenes or elegantly clad ladies copied from fashion magazines.

In Africa the use of ivory was limited, although the material was highly prized and the carvers were very skilled. The African country most famous for its use of ivory was Benin. The ruler, or Oba, of Benin owned every tusk taken from an elephant. It seems that it was the custom of the Oba to keep one tusk from each pair, and sell the other. The money went to the guild of carvers that the Oba supported. The work produced was intricate and beautiful, and much of it was for the royal court. Many of the very old African carvings still in existence today are from Benin, and date back 400 or 500 years.

In Nigeria wealth was displayed in quantities of ivory jewellery in the form of simple bracelets and anklets, and ornaments for the home. In other African countries ivory was used for much less ostentatious items, such as hairpins or ear studs. With the arrival of the Europeans a market opened up for exporting carved objects, but most of the material was exported in its raw state. In the time of slavery it became part of the white gold/black gold story. Local men were commandeered to carry the tusks to the coast and onto the ships, after which the men themselves were ordered to stay on board and were taken away to be traded as slaves.

Arctic ivory has always been looked upon by the local communities in parts of Canada, Alaska, Greenland and Siberia as more of a useful commodity, to be fashioned into utensils and weapons and even used as building materials. Three thousand years ago the Ancient Inuit used mostly walrus ivory, as walruses were much easier to catch than whales. Walrus ivory was also more plentiful than wood. The walrus was altogether an immensely useful animal, as it supplied meat, oil for heating from its blubber, hide for clothes or roofs, bone and ivory. It was only in the twentieth century, with the advent of more sophisticated hunting methods, that whale ivory became as much used as walrus ivory.

The ivory and bone were predominately made into utilitarian objects such as harpoons and other tools for hunting, and scrapers for cleaning the prey. Later, small pieces were used as decoration on wooden objects and clothes. For instance, little flat pieces of ivory depicting fishes or seals were nailed onto boats or harpoons. With the lack of other raw materials, much ivory was also made into children's toys. Babies' rattles might consist of a leather thong strung with a few polar bear teeth, or a toy wooden sled could be pulled by carved ivory huskies. Another very important use was talismanic carvings. The 'tupilaks' – which were half animal, half human – guarded against all manner of evils, for example crawling tupilaks guarded against the demons that hid in the whirl of snow following the sleds (Fig. 3.5).

The Inuit customs have continued through the centuries, until quite recently. Some walrus ivory carvings are still being made, as a by-product of the permitted hunting quota that is part of the Inuit cultural heritage. The carvings are usually small animal figures or jewellery. Nowadays these items are mostly made for export.

Apart from Inuit carvings, there is still some ivory being worked in other countries, notably in the Far East. Hippo ivory is not covered by a hunting ban, and can be exported. However, Europe, the United States, and many other countries have total ivory trade bans, which means that there are no imports of raw ivory to those areas, and which also prevents the import of any finished articles made from new ivory.

4 Bone

When writing about organic gem materials it can be difficult to know which to separate and which to group together. Bone and ivory are usually grouped together because they share many of the same properties and uses, but they are physically different. Antler, on the other hand, is really a form of fast-growing bone. In an attempt to simplify things, each material is dealt with separately in this book. However, there will inevitably be some overlap in the chapters about the three materials.

Structure and properties

Bone is the hard material that forms the skeleton of most vertebrate animals. It consists of a network of collagen fibres impregnated with mineral salts, mostly calcium phosphate. Bone varies in strength and can be as tough as reinforced concrete. Most bones are hollow, the cavity being filled with soft, spongy material. The solid part is interspersed by Haversian canals, which are tiny canals carrying blood, nerves and lymphatics.

The species

Bone for carving can come from any number of sources. In times past even human skulls have been carved. The type of bone most usually used for decorative purposes probably came from cattle, while large items or scrimshaw were made of pan bone (part of the jaw bone), from whales.

Today cattle bone is plentiful and freely obtainable, and sheep bone is also easily obtainable. In Africa trinkets are sold that are made from camel bone. Whale bone is no longer used as whales are now protected species.

Bovids

The most widely used bone comes from animals of the Bovidae family, which includes the domesticated ox – our cattle – and water buffalo. In all probability, it has always been the most popular bone, as it is of a good size and easy to work, though the long bones from the legs – which are the most useful due to their large size – are hollow. Items of bone found in antique shops or seen in museums are also most likely to have been made from an animal in this family.

Whales

Although any bones from a whale could be used, they are mostly not very compact. The exception is the pan bone of the toothed whales of the family Odontocenti (see Chapter 3, 'Ivory'). The pan bone is the flat posterior section of a toothed whale's lower jaw bone. In some species it is very dense. The whole jaw bone of a large whale is the largest single bone in any extant mammal. Whale bone can appear slightly grey or beige, rather than cream coloured.

Other bone

- **Sheep** bone is widely available but is little used today because of its small size. It is also more dense, and therefore more difficult to work, than ox bone.
- **Camels** live in much of Africa, and throughout the Middle East. Their bone has the same properties as ox bone and is used locally. Camels have been introduced to the desert interior of Australia, and also live in Asia as far east as Mongolia, but in these regions their bones are less used.
- **Manatees and dugongs** – sometimes called sea cows – belong to the order Sirenia. They live in the seas around parts of Africa, Australia, Asia, North America, and in the Caribbean. They are not well known and are protected species, so it is very rare to encounter an item made from their bone. However, they should be mentioned because they are unique in that all their bones are compact and not hollow. This has made their bone more versatile for carving than bone from other species, and it has been used as an ivory simulant.

Treatments and uses

Although bone can be put to many of the same decorative uses as ivory, it is less versatile because it is hollow.

Figure 4.1 Small box, showing inserted base and Haversian canals.

Note: Much of the following could equally well apply to ivory (see Chapter 3, 'Ivory').

- Bone can be carved, but only into small or rather flat items due to its thickness.
- Bone can be turned on a lathe and made into, for example, small boxes (Fig. 4.1).
- Bone can take a high polish after carving or other work.
- Bone can be stained to alter its colour. This was frequently done in the manufacture of knife handles.
- Bone can be engraved and inscribed.
- Cut and carved into thin slivers, bone can be used as an inlay material.
- Bone cannot be softened, so must be cut or carved into any required shape.
- Bone is a poor conductor of heat, so has been used for tea or coffee pot handles.
- Bone is more durable than ivory, which tends to warp or crack unless stored carefully.

SIMULANTS

Bone is a relatively inexpensive material and is freely available, so it has not been necessary to imitate it. However, bone has itself been used frequently to imitate ivory. It also bears some resemblance to antler.

- **Plastic.** Due to the availability and low cost of bone, almost the only materials that have been used as simulants – and which could be mistaken for it – are plastics.
- **Reconstituted bone.** This material is not common but does occasionally appear on the market in the form of small figures or netsuke-like 'carvings'. Probably intended to imitate ivory rather than bone, it shows a total lack of structure or marks from carving tools, but will probably show signs of having been moulded. It is made of powdered bone mixed with a filler.

Figure 4.2 Longitudinal section of bone bracelet, showing Haversian canals (magnified).

Figure 4.3 Detail of cross-section of bone bracelet, showing Haversian canals.

TESTS AND IDENTIFICATION

Chemical tests are not used on bone, and identification is by sight and feel.

Visual examination

- Bone lacks the typical identifying features of ivory, such as elephant ivory's 'engine turning', walrus ivory's secondary dentine, or concentric lines in the dentine.
- On a polished, smooth surface of bone, the Haversian canals can be seen. In cross-section these appear as tiny dark dots, while in longitudinal section they show up as straight, thin, dark lines. These are unique to bone (Figs 4.2 and 4.3).
- The Haversian canals will still be visible on a hollow or concave piece of bone.
- Many types of bone and ivory are the same, creamy colour, while antler tends to be darker – pale grey or grey-brown.
- Bone seldom takes as high a polish as ivory and therefore usually lacks the silky feel of ivory.
- The shape of an item may give some indication of its origins (Fig. 4.4). The curvature and hollow structure of bone is limiting, and is

Figure 4.4 Apple corer made from a small bone.

often clearly visible (Figs 4.1 and 4.5). Plastics or reconstituted material have no such limitations.

- Swirls of colour, or the complete lack of structure or Haversian canals, indicate plastic or reconstituted material.
- Air bubbles, marks from a mould, or lack of marks from carving indicate plastic or reconstituted material.
- Plastic is usually lighter than bone, though this may be very difficult to assess in a small or mounted item.
- Some plastics are slightly flexible. Bone is rigid.
- The quality of the work in an item is often the best indication of which material has been used, especially in pieces from the nineteenth and twentieth centuries. Ivory is usually at the top of the scale, followed by bone. Plastics and reconstituted materials come bottom of the list.

Tests

- **Ultraviolet light.** Bone, ivory, and the early plastic casein, display the same chalky blue fluorescence under UV light. All other plastics are inert.

Conservation status and availability

There are innumerable animals that are today protected by trade bans, and bones from these animals are obviously not available for use. However, there is a plentiful supply of bone from domesticated animals.

Past and present uses

Since prehistoric times materials that were readily available, such as horn, bone, ivory and antler, have been put to every conceivable use by humans, including being used as building materials.

Bone is nowadays regarded as far inferior to ivory. Ancient sites such as burial mounds often contain both bone and ivory. Excavation of these sites show that no particular distinction was made between the two materials, regarding either their uses or the craftsmanship involved in working them. It is likely that, in times past, equal value was given to both materials.

Initially, whatever was made had to follow the shape of the material from which it was made. Only as humans learnt to cut and carve the materials did more possibilities present themselves.

Figure 4.5 Modern necklace of sections of bone.

The earliest bone 'carvings' that have been found are reckoned to be about 40 000 years old. They are inevitably primitive, and it is not known what they were used for. In the Stone Age humans discovered how to carve bone to make sewing needles, but before that they made musical wind instruments from the bones of large birds, such as vultures. Some of the 'flutes' found are thought to be up to 24 000 years old. The fashion for bone flutes continued until Roman times and later. The bone most favoured for the instruments was then the shin bones of sheep.

The use of a lathe for bone was a Roman innovation. As well as using it to make turned objects, the Romans used bone to make hairpins, games, combs, spoons and some jewellery such as pendants. It was also used to adorn furniture. It was sometimes stained and coloured, and was polished with beeswax. The Romans in Europe used mostly cattle bone, but antler was also popular and today it is very difficult to distinguish from which of the two materials any given piece was made.

Bone lost its popularity after Roman times, and was seldom used until the Middle Ages, when the work was less intricate than that produced earlier by the Romans.

After the first millennium the emphasis was on ecclesiastical objects, and these were most commonly made of ivory. However, there are examples of bone carvings, both from cattle bone and whale pan bone. Some items were made of part ivory and part bone, which was by then considered a cheaper alternative, and was more easily obtainable.

Ivory and antler were, together with horn, popular materials for stick dressing – the art of making handles for canes and umbrellas. Bone was less used in this format as it is hollow and therefore difficult to carve into a handle, but sections of round, hollow bone have frequently been used on the cane itself as a spacer between the handle and the stick.

In the nineteenth century the whaling fleets of America and Europe produced carvings now known as 'scrimshaw'. A lot of whale bone was used, as well as ivory. The huge pan bones from the toothed whales could be made into large, solid pieces of equipment such as meat hammers, or blocks and cleats for the ship's rigging. Because bone is less likely to warp than ivory, it was also made into more specialised equipment such as rulers and gauges. The smaller, finer and more decorative objects that were produced by the scrimshanders were more likely to be made of ivory than of bone, and were made for use in the home rather than on board ship.

Today, bone is used by local populations all over the world, to make trinkets and souvenirs such as bracelets and pendants for the tourist trade. It is also used to make small, high-quality items. Examples of these are spoons, bowls, beakers, letter openers and napkin rings.

5 Antler

Antler is the fastest growing mammalian tissue, and is a type of bone. Antlers emerge from the frontal bone of the skull of an animal in the mammal family Cervidae. With the exception of reindeer or caribou, antlers are only carried by the males.

Misunderstandings can arise because antler is sometimes called 'stag horn' or 'harts horn' although being osseous (made of bone), it has nothing to do with horn, which is keratinous (made of keratin).

STRUCTURE AND PROPERTIES

Chemically and physically antler is almost the same as bone, though the collagen content is higher. Unlike bone, antler is not hollow, but nor is it completely solid – a compact outer tissue encloses a centre

Figure 5.1 Cross-section of antler, showing core.

which has a spongy appearance (Fig. 5.1). This core permeates the whole antler. The size of the core varies according to the species, and whether or not the animal is domesticated – and therefore castrated – or, in the case of reindeer and caribou, the female has a calf. In these circumstances the core area is larger and the solid parts are thinner.

Antlers grow as pairs, and are incredibly fast growing. They are shed every year after the rutting season, and each year they grow progressively larger and have more branches (called tines or prongs), until a stag reaches maturity. In the case of a European elk (or American moose) the antlers grow over a period of approximately 150 days and can reach a weight of 30 kilograms, which equals almost 1.5 kilograms per week.

Figure 5.2 Antler surfaces: red deer and reindeer lying on elk antler.

Skin or 'velvet' covers the antlers. Beneath the velvet is a mass of blood vessels, which cause the ridged appearance on the exterior of the bony antler material (Fig. 5.2). The velvet cracks and dries when the blood supply is cut off (prior to the antlers being shed), and is rubbed off against shrubs or the ground, which may cause staining to the bony material beneath.

THE SPECIES

Many species of deer grow antlers, and it is possible to gather the material which has been shed. However, nowadays it is also possible to collect antler from farmed and domesticated animals. In some herds they are not left on the animal until they are shed, but are humanely cut off beforehand. This is in order that the velvet can be sold to the oriental medicine trade where it must be used fresh and bloody, before it has died.

Antler is produced by animals in the family Cervidae. That which is most commonly used in Europe comes from red deer, elk and reindeer, though in other countries it is supplied by other, local species.

Cervids are vegetarian. They live on a diet of mineral-rich plants and leaves, rather than grass, as the rapid growth of antler requires the extra nutrients.

Antlers are used for display, status, and for sparring or fighting. The sound of the antlers clashing as two males spar for supremacy in a herd is very impressive, though their encounter follows strict rules and the match is often settled without it actually developing into a fight. Sparring is also a form of play among younger animals.

It is somewhat confusing that the three European animals – red deer, elk and reindeer – have different names in America and are called, respectively, elk, moose and caribou. In fact there are very slight differences between the species on the two continents, for example the American moose is slightly bigger than the European elk. In the following both English and American names are used where it helps to clarify the text, otherwise only the English name is used.

Elk and moose

Elk (or moose), *Alces alces*, live in northern Europe and in Alaska, Canada, north-east USA, and in parts of Siberia and Mongolia. There are six subspecies. The largest males can grow to over 2 metres tall at the shoulder. At maturity they carry huge antlers which can weigh up to 30 kilograms and can have a span of 2 metres. They are branched and palmate, that is to say they branch into points, and their overall shape is wide and flattened, like the palm of a human hand.

Elk are no longer domesticated though they were used as beasts of burden during the Middle Ages. They are hunted for their meat and their antlers are taken as trophies. Shed antlers can be put to the many uses listed later in this chapter, and elk skin makes a very strong leather. Items made from these materials are often found in souvenir shops local to elk habitats.

Reindeer and caribou

Reindeer (or caribou), *Rangifer tarandus*, are best known as the animals that pull Santa Claus' sled. In some countries they are domesticated and do pull sleds. Reindeer live in cold climates: northern Scandinavia, Greenland, northern Russia, Alaska and Canada. Both males and females carry antlers – the only species of antlered animals to do so – those of the males being larger and with more points. The antlers tend to become palmate at the ends, giving a slight shovel shape. The males shed them between November and April, while the females keep theirs until early summer. Some caribou living in areas where there is plenty of food all year round do not have antlers, encouraging the theory that the females carry them in order to look more like young bulls, which helps when competing for food during the long winter months.

Local human populations hunt reindeer for their meat and fur, as well as for their antlers.

Red deer

Red deer (or elk), *Cervus elephus*, have several subspecies. They live in or near woodland in Europe, Scandinavia, North America, North Africa and parts of northern Asia. In Britain they are recognised as the magnificent creatures that roam the Scottish highlands.

Their antlers are branched and pointed. The maturity of an animal can be read in the number of tines, which also gives it status in the herd. A young bull with small antlers would never fight for superiority over a well-established, adult bull, nor would he attract the females. The antlers are at their largest when the animal is between seven and ten years of age.

Wild red deer are hunted for their antlers, which are regarded as trophies and mounted whole, sometimes with the head of the animal still attached. Shed antlers are collected and sold to workshops or factories that work the material. Red deer are now being farmed in some countries. This is mostly for their meat, though the antlers, velvet and hide are also put to use. For example, although not indigenous to New Zealand, red deer have been introduced there and are farmed for meat. As a by-product, the velvet is sold for use in oriental medicines.

Other cervids

Other species of deer live in various countries throughout the world, from the muntjacs in Asia to the little brockets of South America and the tiny pudus of the Andes. Some have impressive antlers while others have just a pair of simple prongs. The cervids living in warmer countries usually have smaller antlers than their cold climate cousins.

Roe deer, *Capreolus capreolus*, are common in much of Europe, North Africa and China. They live in forests, woodland or moorland. In the more populated areas they are found in parks, and often trespass into nearby private gardens to eat the flowers. They are also the deer that are most commonly hunted for sport. They carry simple, branched antlers. Also well known are the fallow deer, *Dama dama*, with their distinctive fawn coats with white spots. They have branched and palmate antlers.

TREATMENT AND USES

The uses of antler are limited because of its shape, the core, and the ridged surface. A whole section of the material is often used to make a single item, leaving the centre and the surface intact as part of the decoration.

- Antler can be carved or turned on a lathe. Sometimes the core is removed, giving a hollow material, but it can also be carved leaving the centre intact. The centre does not take intricate carving, so, if the core is left in place, the carving is usually concentrated around the solid outer part of the material (Fig. 5.3).
- The outer surface of antler takes a reasonable polish, and if the ridged surface is removed, the solid, outer area takes a good polish. The core does not take a good polish (Fig. 5.3).
- Antler can be dyed and stained, but this is seldom done as the colour of the natural material is attractive.
- Although antler can be etched and inscribed, this is seldom done as so much preparatory work would be needed on the surface.
- Antler cannot be softened or moulded and must be cut or carved into the required shape.
- Antler is a poor conductor of heat and has therefore been used as handles for such items as tea or coffee pots.
- Antler is most commonly used for handles, especially for knives. The material is left intact to give a rustic appearance and a good grip (Fig. 5.4). Only good-quality antler can be used for this, as material with too large a core in the centre will not hold the tang (the metal extension from the blade) of the knife. For this

Figure 5.3 Tupilak of reindeer antler.

reason, antler from wild deer is more suitable than that from farmed deer.

Simulants

Antler is readily available and relatively inexpensive, and therefore is seldom imitated.

- **Plastic.** The only material likely to be found imitating antler is plastic, though this is very rarely seen.
- **Bone.** This would not be used as a simulant, but if the core has been removed from antler, it resembles bone, and may cause some confusion. For example, the two materials were used equally in Roman times, and because they are both osseous, it can be difficult, with such old artefacts, to tell the two materials apart.

Figure 5.4 Shoehorn and cigar cutter with handles of red deer antler, and cork screw of polished reindeer antler.

TESTS AND IDENTIFICATION

Antler is easily recognisable by its structure. Even when the ridged outer surface has been removed, the spongy looking core is unmistakable (Fig. 5.1). Only if the centre is absent can there be difficulties.

Visual examination

- Antler is usually a darker shade than the creamy colour of bone. Antler tends to be beige or pale grey, depending on its origins.
- A ridged outer surface or a spongy looking centre is indicative of antler and is easily seen with the naked eye (Figs 5.2 and 5.1).
- Under magnification the polished surface of antler shows a more mottled appearance with lots of tiny, dark spots, and it lacks the straight, dark lines of the Haversian canals in bone. In cross-section antler lacks the small, black dots of these canals.

- A piece of antler with the core removed will usually show some signs of drilling or carving.
- Swirls of colour, air bubbles, signs of moulding or lack of marks from carving tools all indicate a reconstituted material or a plastic.
- Plastics are usually lighter than antler. This may be impossible to assess if the item is mounted or very small.
- Antler is totally rigid, while plastics can be slightly flexible.

Tests

- **Ultraviolet light.** Antler shows less fluorescence than bone under UV light.

Conservation status and availability

Worldwide there are various species of deer that have been hunted almost to extinction, and some tropical and swamp deer are now threatened by loss of habitat. There are also subspecies of red deer that are on the endangered list, but, by and large, antler is plentiful and free from any conservation issues. However, there is a question of ownership, and it is not necessarily permitted for the general public to collect shed antlers, even in apparently public areas.

A past danger to deer was from poachers who sought to obtain the velvet, as this could be sold for use in oriental medicine. Today the risk is greatly diminished as the velvet can be supplied from farmed animals.

Past and present uses

Antler must have been regarded by early man as a gift from the gods. The hunters didn't even have to kill the animal to get it – it just fell off of its own accord. Palmate antlers were probably then more useful than simple, branched ones, but all the material was undoubtedly put to some sort of use.

Antler is an incredibly strong material. There is evidence that antlers were being used as pick-axes 6000 years ago. A length of antler was cut to include the brow tine (the first tine which points forward), and the base of the antler formed the handle while the tine formed the pick. The material was also used to make hunting tools, such as harpoons.

For many thousands of years, humans in some countries have worn antlers for ceremonial occasions. This could be achieved by slicing off the entire top of the skull, complete with antlers, from an animal – usually a red deer – and using it as a hat.

In Roman Europe, bone and antler were used side by side. They were turned on a lathe or carved, and made into all sorts of small items such as spoons or hairpins. Two thousand years later it is difficult to tell whether a piece found at an excavation site is made from antler or bone. The antler used came mostly from red deer, and though the deer were, in later centuries, killed for their meat and the antlers taken as a secondary item, in Roman times only shed antler was used and the meat was not eaten.

In more recent times, while bone lost favour and became a cheap substitute for ivory, antler became more valued in its own right. The attractive, ridged, outer surface of the material was usually incorporated in, or used as, part of the decoration. With the inner core hollowed out, antler was made into powder horns, or pieces of it were mounted on gun butts, where they gave a good grip at the same time as being decorative.

Around the seventeenth century the royal courts of Europe amassed great collections of fine objects, from tiny, painted miniatures to massive, ornate, centrepieces for banqueting tables. Organic materials of every description were incorporated and many magnificent examples can be seen in museums today. A great favourite was the so-called drinking cup – which may have been ceremonial, but was more likely to be ornamental – from which it would have been almost impossible to drink. In the case of antler, a silver gilt cup could be mounted on a whole antler, supported by more ornate metalwork. Another way to use antler was to encrust an object such as a tankard with slivers of the material.

Along with many other organic materials, antler was used to make ornate handles for walking canes and umbrellas. Here again a piece of antler would be left in its natural state displaying the ridged surface, and polished. This was mostly due to its own, attractive appearance, but also because of the fact that antler cannot be shaped by pressing or moulding.

The most enduring use for antler over the centuries has probably been as knife handles, where the end of the antler tine is used in its entirety and simply polished. Knives such as these have been popular in all sizes, from small pocket knives to hunting knives. The material sits well in the hand and the ridged surface gives a good grip (Fig. 5.4).

In the north, the Inuit regarded antler in much the same way as they regarded bone and ivory. The materials existed to be utilised, and little went to waste. The meat from the animal was eaten, the skin tanned to make leather, fur provided clothes and bedclothes, and the bones, teeth and antlers were carved into anything from harpoon points and amulets to children's toys.

Today, handles made of antler tines are still very popular, especially for knives. The best material is from red deer, as their antlers have so many tines.

6 Rhino horn

Rhino horn comes from the rhinoceros, of the family Rhinocerotidae. There are five species of rhino extant today. Three of the species carry two horns in tandem on the nose, the horn at the front usually being larger. The remaining two species carry only one horn on the end of the nose.

Structure and properties

Horn from the rhino is different from other types of horn, as it is composed entirely of compacted strands of keratin, so that it is solid with no bony core. Other types of horn are thin sheaths of horny material covering a bony core. Rhinos' horns grow throughout the animals' lives, and can regrow if broken or damaged. They grow from the skin on the animals' noses (Fig. 6.1).

The species

Rhino used to live worldwide but now only inhabit Africa and parts of Asia. There are five species left, all of them severely under threat of extinction. Their horns are all similar in shape and colour, curving very slightly from a broad base to a pointed tip. The horns are used partly for fighting, but mostly for foraging.

The **African white rhino**, *Ceratotherium simum*, is an animal second in size only to the elephant. It inhabits the African savannahs and has two horns (Fig. 6.1). The larger one in front has been known to reach a length of 1.5 metres.

The **African black rhino**, *Diceros bicornis*, is smaller, and also bears two horns. It inhabits forests and scrubland.

The **Indian rhino**, *Rhinoceros unicornis*, has, as its name suggests, only one horn. It is found in India, Nepal and Bhutan. It is different

Figure 6.1 African white rhinoceros (Suzie, at London Zoo).

from the other rhinos in having a skin that appears to be in folds or sections and looks a little like armour plating.

The **Javan rhino**, *Rhinoceros sondaicus*, is similar to the Indian rhino but somewhat smaller and with a smaller horn. It lacks the distinctive folds in the skin.

The **Sumatran rhino**, *Dicerorhinus sumatrensis*, is the smallest of them all. It has two horns.

Treatments and uses

Rhino horn consists basically of keratin, as do other varieties of horn, and therefore, in theory, could be treated in many of the same ways as other horn and be moulded or pressed to a desired shape. In practice, this has not happened, due to the very high value of rhino horn, and because it is solid.

- Rhino horn can be carved and turned.
- Rhino horn takes a good polish.
- It is possible to make very intricate, three-dimensional carvings in rhino horn. This work was common on old rhino horn items such as libation and drinking cups, and dagger handles (Fig. 6.3).
- Rhino horn can be painted or inlaid.

- Old rhino horn objects have often been mounted with silver or silver-gilt.
- In powdered form rhino horn is used as an ingredient in medicines, notably to neutralise poison and to bring down fevers.

Simulants

Rhino horn imitations are extremely rare.

- **Plastics** could imitate rhino horn. They are unlikely to be convincing due to their lack of structure.

Tests and identification

Visual examination

- At first glance rhino horn has an even, mid-brown colour, and looks a little like wood, or even plastic. Closer inspection reveals that it has a striped brown pattern, following the lines of the tubular formation of the material. It can be very slightly paler on the outside surfaces, though unlike cattle horn, it does not display distinct colour variations or patches of colour (Fig. 6.3).

Figure 6.2 Structure of rhino horn. Magnified.

- Under magnification, rhino horn displays a very fine, tube-like pattern, appearing as lines in longitudinal section, and a compact pattern of tubes in cross-section (Fig. 6.2).
- In thin sections of a carving, rhino horn can appear almost translucent.
- The type of item under examination may give some indication as to the material used. For example, rhino horn was often used to make ornamental drinking cups. Intricate workmanship is also indicative of a valuable material (Fig. 6.3).
- Lack of marks from the carving tools would indicate plastic, though age and polishing can remove tool marks on the most prominent surfaces.
- Plastic imitations may show marks from the moulding process, and possibly air bubbles.
- Unless weighted, plastic is lighter than rhino horn. Some woods are of a similar weight to the horn; however, wood has a different and slightly drier feel.
- Plastics would display a lack of tubular structure.
- Very even colour, or swirls of colour, would indicate plastic.

Tests

There are no chemical tests for use on rhino horn, and very little else that can help with its identification. However, the material is so unique that it should not be necessary to undertake more than a visual examination.

- **Sectility.** This is destructive and cannot be recommended. Besides, it is not a useful test as both plastic and horn can also be pared with a sharp knife.
- **Burning.** Any result would not be conclusive, as rhino horn will burn with the same smell as other horn types. This is also a destructive test and is not recommended.
- **Ultraviolet.** Rhino horn reacts very little under UV light. The slightly paler outer surface of the horn may fluoresce slightly with a chalky blue colour.

Conservation status and availability

All species of rhino are endangered and some are almost extinct, for example there are probably fewer than 100 Javan rhinos left today. Most rhino species are listed on CITES Appendix I, with only one subspecies on Appendix II, for very specific reasons and under the strictest control.

These rules are generally accepted worldwide, though there is still some poaching of rhinos, as there are always people who will pay for their illegal slaughter in order to obtain the much-prized horn.

The Yemeni market was one of the largest for this material, since owning a ceremonial dagger with a handle made of rhino horn was a strong tradition, and all young men of status had to have one. A large amount of horn went to make one handle – often a whole horn was used. This trade was stopped voluntarily in 1997 in order to spare the animals, some years before the Yemen signed the CITES agreement.

Rhino horn is still much in demand as an ingredient in Far Eastern medicines, notably in China. Powdered rhino horn has a value that far exceeds that of either cocaine or gold. The popular misconception is that rhino horn is used solely as an aphrodisiac. In truth, that is only one of its uses, as it is an ingredient in medicines for various ailments. One notable use is as a medicine to reduce fevers in children, which makes the continued demand for the powdered horn rather more understandable.

Poaching is not the only threat to the rhino: loss of habitat also plays a large role. Ironically, the conservation of elephants in Africa has exacerbated the problem for the black rhino as they live in forests and scrubland, and elephants, confined to certain areas for their own protection, are felling the trees which made up the black rhino's habitat. Mankind is also to blame for clearing these areas in an effort to eradicate the tsetse fly which can infect domestic animals. In some countries, civil unrest has also caused conservation programmes to be abandoned.

When animal numbers get low, the natural mortality rate may overtake the birth rate and the creature will die out, no matter how well the hunting bans have been observed. All species of rhino are therefore considered to be at risk, and it is unlikely that rhino horn will again be available to buy.

Past and present uses

Rhino horn has a different history to that of other horn types. African rhinos were unknown until the mid-nineteenth century, but Indian rhinos were tamed and used to pull ploughs. They were also sent into battle where it was believed that their skin was bulletproof, because it was thick and hung in folds.

Although it cannot stop bullets, rhino hide is incredibly thick and strong and can deflect arrows, so it was used for shields. Being thick and pliable, rhino skin shields could be painted and decorated, as they were in India, or they could be polished and adorned with silver, as was the custom in Abyssinia.

Figure 6.3 Rhino horn cup. Chinese. Eighteenth century. *Victoria and Albert Museum, London.*

Rhino horn has been used for centuries, by many nationalities, for a variety of purposes. The trade to China goes back 2000 years. For half of this time it was customary for Chinese aristocrats to present their emperor with rhino horn vessels to celebrate a birthday. These could be in forms such as cups, bowls or brush pots, and were intended to be treasured, not used (Fig. 6.3). The Chinese greatly admired rhino horn and, with their usual beautiful workmanship, carved it into all manner of items, from buttons and belt buckles to combs, bracelets and talismans.

In many parts of the world, including Christian, Buddhist, Hindu and Muslim countries, drinking cups made of rhino horn were thought to be able to detect poison, which was supposed to bubble in the cup. In Japan it was carved into netsuke, and in Europe, in the early twentieth century, it was being used for everything from pistol grips to door handles. In Africa rhino horn was always used for far more practical purposes, such as utensils, or as an ingredient in some simple medicines.

It was in the Far East, especially in China, that belief in the medicinal power of rhinoceros horn was so great. In sixteenth century China, imported horn from India was being credited with the ability to cure snake bites, hallucinations, typhoid, headache, food poisoning, boils and fever. Contrary to western popular belief, there was very little indication that it would act as an aphrodisiac.

7 Horn

Horn is the most versatile of all the organic decorative materials. It has been put to utilitarian use for thousands of years, and has been used as a gem and decorative material for centuries.

Horned animals are plentiful and occur worldwide in a huge variety of species. Cows, sheep and goats live among us as domesticated animals giving milk, wool and meat. In India, cows roam the streets and are regarded as sacred. Antelope live in the bush, the tropics, and the mountains, and buffalo and bison graze in the plains. They all belong to the family Bovidae.

In much of the western world today our cattle do not carry horns. This is because they have been de-horned – a simple operation carried out by cauterisation – in order to prevent the cattle injuring each other. In the natural order of things the males of each species carry a pair of horns which are a status symbol and are used for sparring. In most species the females also carry horns. These are usually a little smaller and thicker than those of the males.

The size and shape of horns are as varied as the animals that grow them. They can be short, fat, long, thin, spiral, curve, corkscrew, and point in almost any direction. The African impala has beautiful lyre-shaped horns, while the Arabian oryx has long, slender, straight horns that point upwards and away from its face. The African great kudu has large horns that corkscrew, and the African buffalo's head is framed by heavy horns that form a solid horn plate across its forehead, before curving down on either side of its face.

Structure and properties

Horn from the Bovidae family is made of hard keratin formed in laminated layers. A horn varies in thickness according to the species and age of the animal. Usually only the tip is solid, while the rest of the horn forms a sheath covering a bony core, which is attached to the frontal bones of the animal's skull. Horns are permanent structures, which grow continuously throughout the animal's life.

The colour in horn comes from the pigment melanin. The colour varies from cream through browns to black, or can be a mixture of these colours, depending on the species.

All types of horn are waterproof and greaseproof, and can be made into items that are airtight. The material is also thermoplastic so can be heated and moulded, retaining its shape after cooling.

Note: Baleen (see Chapter 12, 'Miscellaneous organics') and hoof are very similar to horn and have in past times been put to many of the same uses. They are also made of keratin, but are solid. Like horn, hoof also consists of laminated layers of keratin. The structure of baleen is different: it consists of very fine tubes of keratin, held together by a coating of the same material.

It can be difficult, if not impossible, to tell whether a finished item has been made from hoof, baleen or horn. The shape of the finished item may give some indication as to its origin, as the natural shape will, to some extent, influence how the material is fashioned.

Derived from whales, baleen is no longer used today. Hoof is used mostly for the production of gelatine or fertiliser.

The species

Any horn can be used to make items of jewellery, boxes, and countless other objects, and there are examples of weird and wonderful items

Figure 7.1 Water buffalo in Vietnam.

in many museums. Throughout the world people have used whichever variety of horn is most easily available. Only a few types are much used today and most of the horn used in past times probably came from the same, or closely related, species.

Water buffalo

Native of South East Asia, the water buffalo, *Bubalis bubalis* (Fig. 7.1), is domesticated and used as a working animal, which can pull a plough or turn a wheel for an irrigation system. It is farmed for its meat, and its hide is turned into leather.

Both males and females carry horns, which are long, slightly curved and elliptical. The inside of the curve, close to the base, is ridged. They are usually an almost even, dark brown to black in colour, though the horns of a young animal can be paler and show more colour variation (Fig. 7.2). The horns are mostly hollow with a solid tip, but are thicker than ox horn.

Some of the raw material is exported, but much of it is made into horn objects such as bowls and salad servers, before it is exported. Tourist souvenirs made of horn are readily available locally in South East Asia. A common example is a pair of horns, which are left whole, polished, etched with pictures of dragons or similar, and mounted on wooden bases for display.

Ox

The most widely used horn type is from domesticated cattle of the genus *Bos*. In England some of our cattle are polled, that is to say they are bred hornless, but most are de-horned. Some species still carry horns, for example the French charolais, which has small, pale horns that are easy to work and give beautiful results.

Cattle from further afield often carry much larger horns. The variety most used in the UK today comes from Nigeria, from the domesticated oxen of this region, *Bos indicus*. Mature horns from this animal are about 80 centimetres long and vary in colour, often pale cream to beige at the base and becoming almost black at the tip. Ox horn is more round than that of the buffalo, more hollow and somewhat thinner, but it is smooth and has no ridging. The horns taper in thickness from a solid tip to around a quarter of a centimetre at the base. These horns take a very high polish and display an attractive colour pattern.

Ram

Rams' horn used in the UK comes from various breeds of sheep belonging to the genus *Ovis*, among them Herdwicks and Swaledale.

Figure 7.2 Carving of bird. Thailand. Young water buffalo horn.

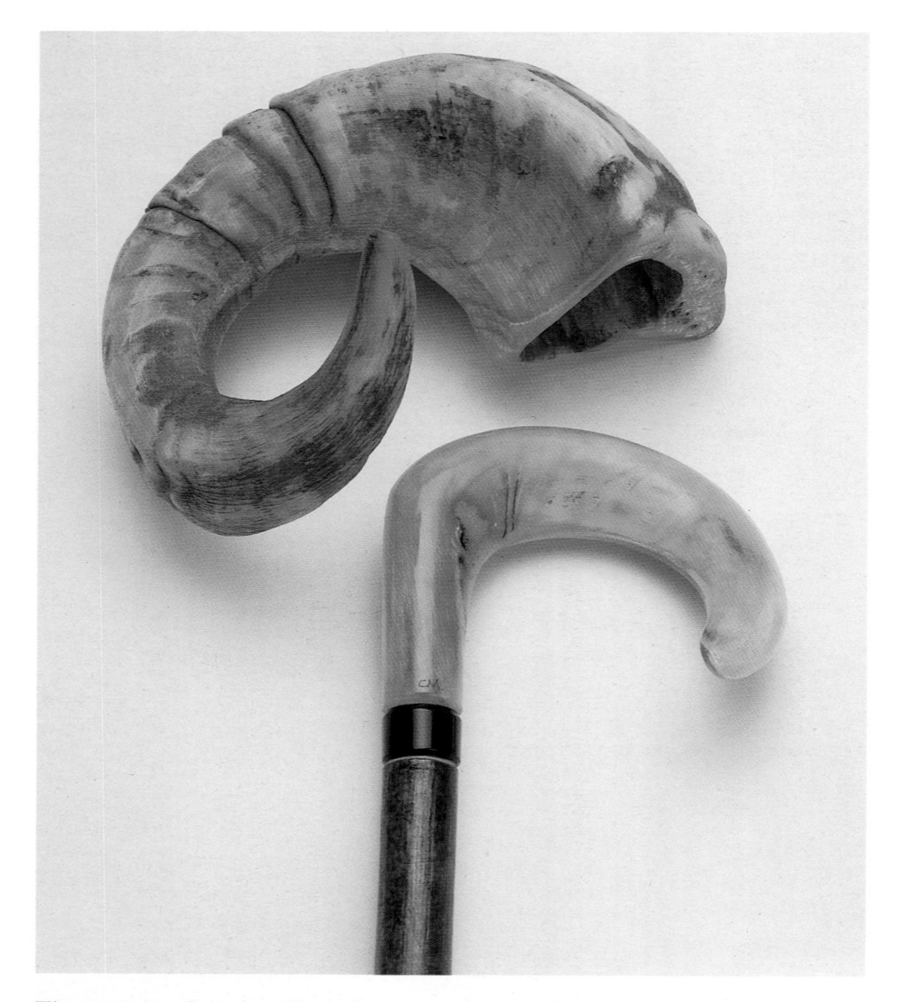

Figure 7.3 Cane handle and untreated horn. Both ram's horn.

Rams' horn is more compact than other types of horn, so it is harder to work, but gives very beautiful results. The sheath varies in thickness around its circumference, and is almost triangular in cross-section. It is also curved and ridged (Fig. 7.3).

Little rams' horn is used today because of its complicated shape. When worked it is usually compressed into a solid piece of material which is best used for cane and stick handles. Polished rams' horn varies in colour according to the species, but is mostly a uniform colour, often a very attractive, silky-looking pale honey (Fig. 7.3). UK supplies today come from domestic herds in England and Scotland.

Figure 7.4 Various modern horn items.

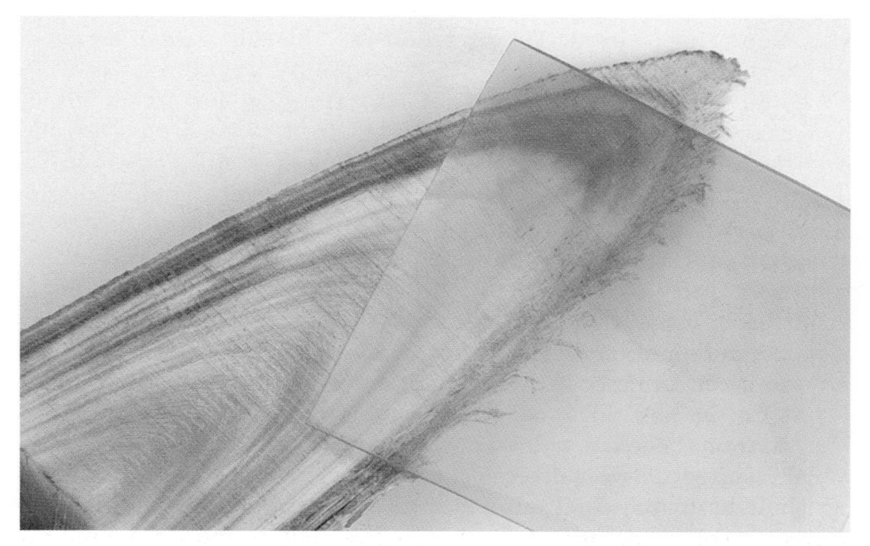

Figure 7.5 Piece of 'lanthorn' over a section of unworked horn tip (magnified).

TREATMENTS AND USES

- The natural surface of horn is dull but the material takes a good polish. The better the material, the higher the polish it will take.
- Being thermoplastic, horn can be heated and moulded under pressure. This can be either a dry or a wet process (Fig. 7.4).
- In the dry process, the horn is softened by heating over a flame, then flattened in a heated press, reheated by dipping in hot oil, and finally shaped in another press. The heating is very critical as too much heat will turn the horn brittle, while if it is not heated enough it will revert to its original shape.
- Today the wet process is only used when making clear 'lanthorn', which was once used in place of glass in lanterns (Fig. 7.5). In this process, pale horn is soaked in water for several weeks before being boiled, again in water, to release sulphur and calcium to clarify the material. It is finally flattened in a heated press. The resulting clear material may be called 'green horn', but this term has several meanings.
- In past times, pieces of horn were heated, probably in tallow, and welded together edge to edge, to produce larger pieces.
- The bases of, for example, drinking beakers are today glued in place using modern glues. Formerly, a glue made from horn or hoof would have been used.
- In past times horn could be peeled into thin layers to use, for example, as a veneer in place of tortoiseshell. This is not done today as modern horn does not peel so easily.
- Horn can be carved. The part most used for this is the solid tip of the horn.
- Horn has been widely used for stick dressing, that is the art of making a decorative – often carved – handle or knob on a walking stick or cane. Curved rams' horn lends itself to this craft as it has approximately the right shape and needs only to be compressed by dry heat and pressure to make it solid (Fig. 7.3).
- Horn can be turned on a lathe to make, for example, drinking beakers.
- In the past, hoof was used in the button industry. The hoof was cut into strips and the round button blanks punched out. Today hoof is little used for decorative items.
- Horn can be embossed with a pattern or a picture on the surface, as can be seen on some ornate buttons and boxes (Fig. 7.6).
- Horn can be inlaid with silver 'piqué' work or with mother-of-pearl. As with tortoiseshell, this is done under heat and pressure, and the material to be inlaid is pressed into place. It is not necessary to carve out a hollow for the inlay. Theoretically, it should not be necessary to use glue as the material will contract to hold the inlay as it cools, but in practice, glue is often used.

Figure 7.6 Pressed horn button. Late nineteenth century (magnified).

- Today powdered horn is mixed with a filler to seal cracks or cover blemishes in horn items such as cane handles.
- Horn can be bleached to a pale opaque colour with an almost iridescent sheen. This was seen in art nouveau jewellery (Fig. 7.13).
- Horn can be dyed, as a rule with organic dyes. The dyed colour is only on the surface. In thin layers dyed horn could be used as coloured veneer on furniture.
- Horn was dyed black as a cheap imitation of jet, when this material was popular for jewellery.
- Horn can be bleached and painted, as, for example, when imitating tortoiseshell. This was sold as 'mockshell' in the nineteenth century and was used mostly for hair ornaments, fans and combs.
- Horn does not last forever, but deteriorates with age. For this reason there are few very old items to be found in museums. Damp, heated storage is especially damaging to horn.
- The patina of horn is retained by handling it, hence it is a good material for use in stick dressing.
- Horn is not affected by salt or vinegar, so it is ideal for use as salt spoons and salad servers.

- Horn is affected by hot water and detergents, and will revert to its original shape if washed in a dishwasher.
- Scrap from the production of horn objects, and hoof, have been used to make glues. These have largely been superseded by modern, inorganic glues.
- Similarly, horn and hoof have been used to make fertiliser. Health regulations in some countries nowadays prevent this from being economically viable.

SIMULANTS

There have been few real simulants of horn as supply is plentiful and the material is relatively inexpensive. However, it is not cheap to work as it must be hand crafted, so today it is regarded as a luxury item. Horn has itself been used as a simulant of other materials, most notably tortoiseshell, black coral and jet.

- **Reconstituted** horn made from powdered horn with a filler could perhaps be confusing if it has been dyed and pressed or moulded, as natural horn will also lose much of its striated appearance when moulded. The process of reconstituting horn to make whole objects has been little used as it has not been very successful, but the substance is used to fill in cracks or blemishes on natural horn items.
- **Plastics** are the one, obvious imitation of horn. Today many of the objects that we associate with horn, such as shoehorns, are made of modern plastic and have been given a pattern imitating the natural material. Plastic is more readily available and far less costly to manufacture.

TESTS AND IDENTIFICATION

It is unlikely to be necessary to test horn, as visual examination should be enough.

Visual examination

- Striations in horn that are due to its structure will usually be visible to the naked eye, though they can be faint in material that has been moulded or clarified, such as lanthorn. They may be more clearly seen under magnification (Fig. 7.7).
- The colour variation tends to loosely follow the striations (Fig. 7.7).
- Patches of minute blobs of brown colour, together with a lack of striations, indicate that the material may be tortoiseshell (Fig. 8.15).

Figure 7.7 Horn structure (magnified).

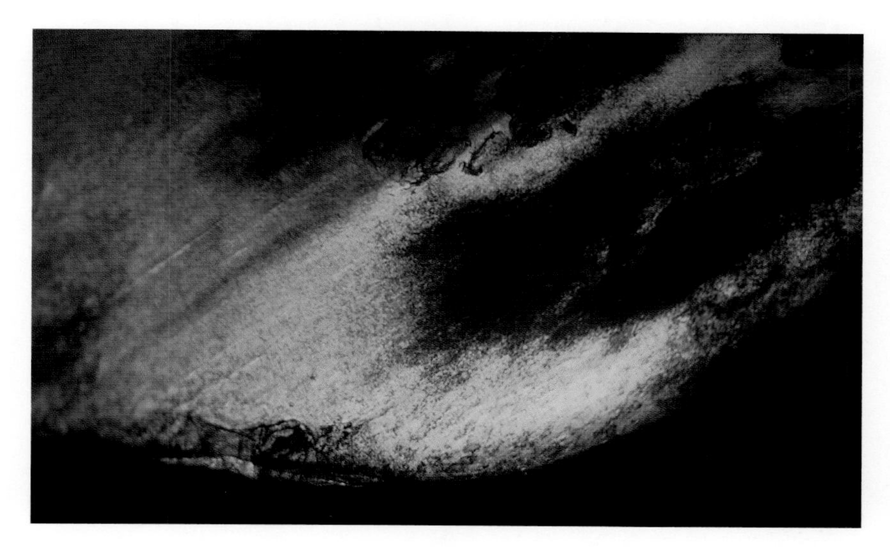

Figure 7.8 Horn structure under crossed-polars (magnified).

- In cross-section, for example in buttons carved from the tip of the horn, the fibrous structure has a furry appearance, possibly with rings of differing colour (Fig. 7.9).
- In plastic imitations, there is a complete lack of structure, and any colour variation will appear as random swirls or blobs (Fig. 7.10).
- It may be possible to see air bubbles in plastic imitations. They do not occur in natural horn.

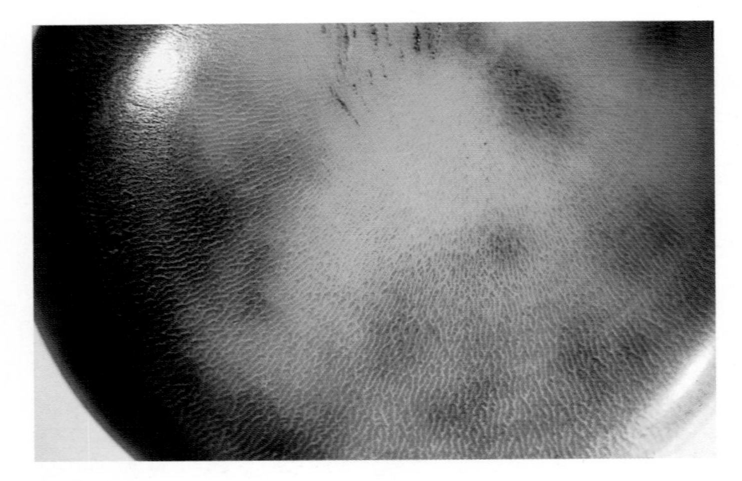

Figure 7.9 Horn button, showing structure.

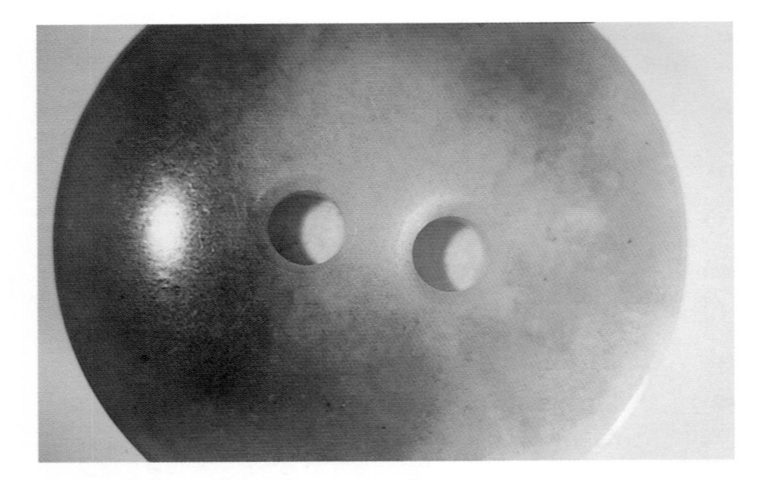

Figure 7.10 Plastic button, showing swirls of colour.

- Pressed or embossed horn has soft and rounded edges (Fig. 7.6).
- Carved horn may show marks from the carving tools. Edges and corners will be much sharper than with pressed horn.
- Surface deterioration is not a reliable indication of the nature of the material as both plastic and horn dull with age and both scratch easily.
- Piqué work, good workmanship, or a good finish would all be indicators that an item is made from horn or tortoiseshell, rather than plastic.

Tests

- **Ultraviolet light.** Viewed under UV light both horn and tortoiseshell show a slight fluorescence. This is more noticeable in the paler areas of the two materials.
- **Crossed-polars.** This is a difficult test to carry out as horn is often not sufficiently transparent to allow light to pass through it efficiently, and because many pieces are too large to put onto a polariscope. Where it is possible to use this method of testing, and with the added help of magnification, horn will show multicoloured lines and spots (Fig. 7.8), whereas plastic is inert.
- **Sectility.** This is not a useful test as horn, like tortoiseshell and plastic, can be paired with a sharp knife.
- **Burning and hot-point tests.** Horn burns at a higher temperature than tortoiseshell, but, also being made of keratin, gives off the same smell of burning hair. Plastic burns with an acrid smell. This is a destructive test, but if it must be carried out, it is advisable to pair off a tiny section from an area where it will not be noticed and burn this. Inserting a hot needle into an object risks melting a large hole, and the object may combust.

Note: One of the early plastics, which was used to imitate horn is celluloid. *It is very highly inflammable* and can explode. Extreme caution is urged when attempting the burning test. By testing only a sliver of the material the danger is lessened, but it must be emphasised that if there is any suspicion that the material might be celluloid, this test should be avoided.

Conservation status and availability

Very many animals that carry horns are protected, as they have been overhunted. An example is the Saiga antelope, which lives in northern Asia and has horns that are popular for use in oriental medicine. Many of the African horned animals are nowadays also protected. However, the types used today by the horn industry are plentiful and are freely available. Most of the material comes from domesticated animals, bred in captivity. The importation of horn into Europe is very strictly controlled and the material that arrives is cleansed and disinfected. Occasionally, horn from a particular source cannot be obtained because of a changing political situation in the country of export, or because of illness, such as foot and mouth, which stops all movement of animal products.

Past and present uses

Horn deteriorates and finally disintegrates with age, so there are no examples of its very early use. However, it is possible to detect traces of the material in, for example, burial mounds, using modern, high-tech equipment.

Without doubt it has been put to use by humans ever since earliest times. In its raw state it made a good, waterproof container. It could be

Figure 7.11 Modern horn necklace.

used as a scoop or a drinking vessel, and large horns could be used as building materials. In many parts of the world, horns were added to primitive, wooden hunting hats and masks to give the wearer a fierce and dangerous appearance, which supposedly helped to frighten the prey into submission – the bigger the horns, the fiercer must be the hunter.

The Romans used pale horn, split and flattened into thin layers, as translucent window panes. For centuries these thin sheets of 'lanthorn' protected lanterns from draughts while still allowing the light from a candle to shine through (Fig. 7.5).

At some stage it was discovered that horn could be softened and bent by heating or boiling it. Further, bits of horn would stick together when they were heated, and glue could be made from the material. This opened the way to a multitude of uses, such as spoons and ladles, or drinking cups with a watertight base that could stand on a flat surface.

Contrary to popular belief the Vikings did not adorn their helmets with horns, but they did make drinking cups from the material. The metal mounts for these have been found in Viking burial mounds, though the horn itself is long gone.

Large, mounted drinking horns were used for ceremonial occasions, and everyone present would drink from the same horn as part of the

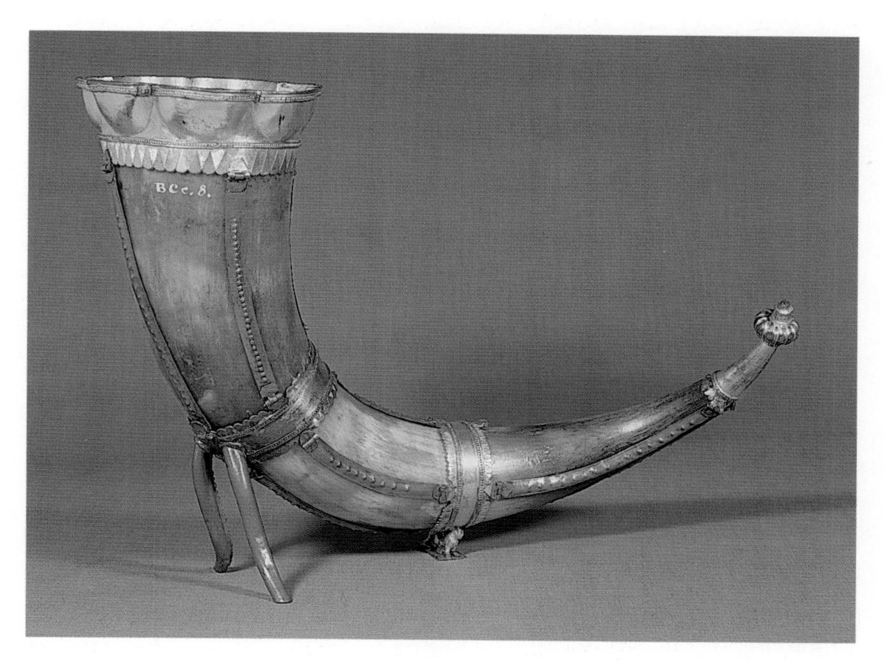

Figure 7.12 Ceremonial drinking horn. Fourteenth or fifteenth century. *The National Museum of Denmark.*

ritual. Silver and bronze mounts survive from classic times, and from many countries, but the earliest surviving horn material probably only dates from the mid-fourteenth century. Complete mounted horns dating from the fifteenth and sixteenth centuries can be seen in many museums (Fig. 7.12).

Making horn objects was not a cottage industry. By the year 1284 there was a horners' guild in London, which still exists today, and which is now called the Worshipful Company of Horners. When first formed, they controlled the sale and purchase of horn within a radius of 24 miles of London. They protected the horners' rights and their welfare, and they kept an eye on the quality of the work produced. As the years passed they amalgamated with the leather bottle makers, and, finally, in 1943, with the plastics industry – today's equivalent of the horn of the Middle Ages.

In many museums there are collections of old, beautifully decorated rifles and blunderbusses, and hanging alongside are the powder horns. Almost invariably made of horn (hence the name), they are naturally curved. The large end is sealed with a horn disc, and the narrow tip – which is ideal for pouring the powder into the narrow opening of the gun – is cut off and replaced with a horn stopper. Some are plain, and some are mounted in silver or otherwise decorated, but the concept remains the same. Horn was not only good for holding powder as it was waterproof and airtight and kept the powder dry, but it had the right shape as well.

Snuff boxes were, on occasion, made from hoof which had been hollowed out and given a hinged lid. They frequently retained the shape of a hoof but could also be fashioned into a less obvious shape. The prettiest snuffboxes, often termed snuff mulls, were made from small, curved rams' horns. The hollow interior of the horn was cleaned and the exterior was polished to a high sheen. An ornate silver stopper, and possibly a silver plaque on the side, added the finishing touches. These are now collectors' items.

Sounding horns have been known for thousands of years in many parts of the world. The horn was cleaned and polished, the rough edge was removed and the tip was cut off to allow one to blow through it. Some may have been embellished with precious metals, while some were left plain and simply polished, all depending on their intended use. More ornate work was applied to horns that were to be sounded for ceremonial and religious occasions, than to those hung on a leather thong and used by shepherds.

By the late fifteenth century, the art of embossing horn had been discovered and countries such as Austria, Germany and Italy were producing horn articles with embossed scenes of an ecclesiastical nature. At this time, baleen was also being used, and today it is not possible to tell which of the two materials was used for any given piece,

Figure 7.13 Horn pendant by Mme Bonté. France, early twentieth century. *Jewellery by kind permission of Didier Antiques, London.*

as, once embossed, their slightly different structures were destroyed.

Making handles for canes and walking sticks or umbrellas is known as stick dressing. This was traditionally a craft carried out by shepherds who carved the horns while minding the sheep. The raw material might come from an animal that had died, or from an animal that had lost a horn in a fight. The stick handles produced were often ornate and beautiful, following the shape of the horn and frequently depicting an animal's head – for example, that of a dog or a lion. Today stick dressing is regarded as a hobby.

The horn industry produced countless items such as knife handles, pocket knives, buttons, boxes, shoehorns, glove stretchers, hair combs, backs for hair brushes, fans and jewellery. Some of it displayed very simple workmanship (Fig. 7.4), while some was intricate and involved skilled craftsmanship.

Horn jewellery is, as a rule, thicker and less delicate than jewellery made of tortoiseshell, though it can be difficult to tell them apart. A notable exception is the art nouveau jewellery, made at the turn of the twentieth century. Somewhat surprisingly, horn was used more than tortoiseshell, and was bleached, carved, and combined with enamel, gold, ivory, pearls and precious stones to produce magnificent pieces of extraordinary beauty. Masters of this art included Mme Bonté (Fig. 7.13) and George Pierre, and of course the ultimate master, René Lalique.

At the other end of the scale was horn furniture. Produced in Germany this became popular in Britain in the mid-nineteenth century, especially for use in hunting lodges. In America there are examples of buffalo horn furniture. These are all very large pieces and were probably intended more as trophies than as items on which to relax.

Today horn items of all descriptions are produced in many countries worldwide, often as souvenirs for tourists. In Africa simple hoops of horn are sold as bangles, while in South East Asia the horn is carved and engraved. Popular subjects are carvings of birds or ships. Whole horns are carved into mythological creatures such as dragons. These items may be intended exclusively for the tourist trade, but the workmanship that goes into their production is usually of a high quality.

In Europe only a few small, horn workshops or factories remain. In all these establishments the work is hand crafted, using much the same methods as were used centuries ago, because horn is not a suitable material for use in mass production by machine.

8 Tortoiseshell

Tortoiseshell is the name given to the mottled, golden or reddish and brown, translucent, horny plates or 'scutes' covering the shells of certain species of marine turtle. The best known and by far the finest tortoiseshell comes from the hawksbill turtle, but the scutes of the green and the loggerhead turtles have also been put to some decorative uses.

'Blond' tortoiseshell comes from the scutes on the belly of the turtle, which are a pale honey colour and almost transparent. They are thin, and several must be welded together in layers for the material to be thick enough to work.

Although marine turtles are constantly being researched, our knowledge of them is somewhat limited as, from within minutes of hatching, they live their entire lives in the sea. Only the females come ashore to lay their eggs, after which they return quickly to the sea (Fig. 8.1). It

Figure 8.1 Female hawksbill turtle returning to the sea after laying eggs. Barbados.

appears that female turtles always attempt to return to their own natal beaches to lay their eggs deep in the sand. Nowadays this can be problematic for the turtle, as so many beaches have changed over recent years due to the rise in tourism.

A marine turtle can lay up to 800 eggs in one season, in batches of 90 to 150. The eggs hatch about nine weeks after they have been laid. The baby turtles (the hatchlings) emerge at night and make their way straight into the sea, attracted by the celestial light shining on the surface of the water.

In the sea, turtles can stay under water for a short period of time, but must surface regularly to breath.

Structure and properties

Tortoiseshell is made of keratin, the same material as, among other things, horn. The colour in tortoiseshell comes from the pigment melanin, deposited in minute globules of about 100 micron in diameter. Like keratin, melanin is very common in the animal kingdom as it is also the pigment which gives, for example, feathers, skin, and octopus 'ink' their colour.

Tortoiseshell is a thermoplastic, that is to say it can be softened by heating, and then shaped or moulded, retaining the new shape when cooled. The process can be repeated.

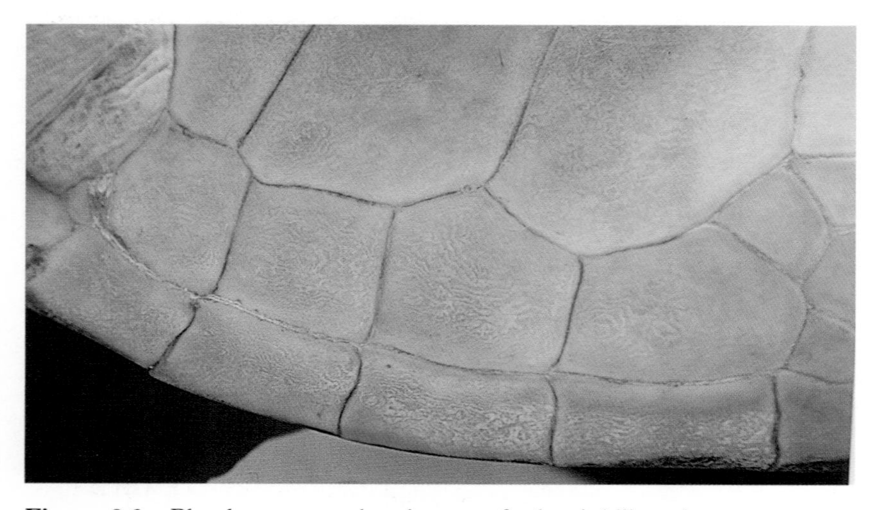

Figure 8.2 Blond scutes on the plastron of a hawksbill turtle.

Figure 8.3 Overlapping scutes on the carapace of a tagged hawksbill turtle.

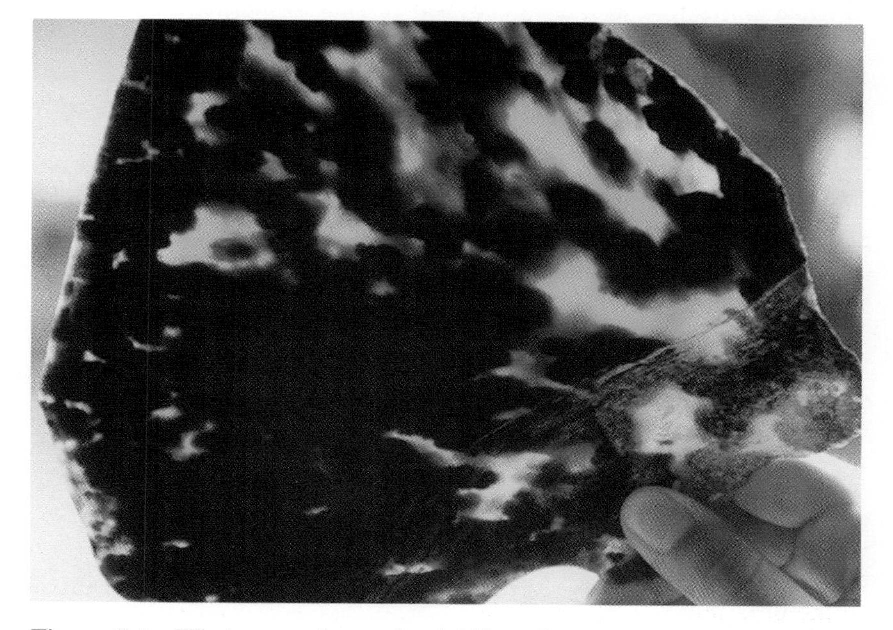

Figure 8.4 Whole scute from a hawksbill turtle.

SPECIES

Marine turtles have bony shells and belong to the family Cheloniidae. Unlike tortoises, the marine turtles have flippers in place of legs, and cannot retract their heads and tails into their shells.

All cheloniids have a shell covered in horny scutes, which grow continuously. The shell on the animal's back is called the carapace, and the shell on its belly is called the plastron. These two shells are each made up of a horny outer layer attached to a bony layer underneath. The bony layer of the carapace is also fused to the animal's vertebral column, plus its shoulders, pelvic girdle and ribs. The scutes on the plastron are typically paler in colour than the carapacial scutes, and they are fully attached to the bony surface beneath (Fig. 8.2). The scutes of the carapace are usually mottled, and, with the sole exception of the carapacial scutes of the hawksbill turtle, they are also fully attached to the bony layer underneath.

Hawksbill turtle

The hawksbill turtles, *Eretmochelys imbracata*, grow the best tortoiseshell. They get their name from the narrow, hooked, beak-like appearance of their mouths. They live around coral reefs but can travel long distances, as has been proved by turtles that have been tagged and tracked using satellite transmitters. Hawksbills are found worldwide in tropical and subtropical waters, and they feed primarily on sponges.

The female is larger than the male, reaching up to almost 100 kilograms, with a carapace measuring up to 95 centimetres. The hawksbill is the only turtle that has overlapping scutes on the carapace. Only the front of each scute is fused to the underlying bony layer, allowing the rear edge to overlap the next scute, like tiles on a roof (Fig. 8.3). Because of this growth habit, the hawksbill is also the only turtle to grow scutes of more than a few millimetres thick – the centre of the main scutes on the carapace can be almost a centimetre thick on a mature hawksbill turtle. On a live animal, the scutes on the carapace appear a dark, dull, opaque brown, and may even be covered in barnacles, but when polished they show the beautiful honey-gold and brown mottling for which the material is famed (Fig. 8.4).

Tortoiseshell derived from the hawksbill is very versatile and can be used in numerous ways, unlike that of the other turtle species, where it is much thinner and so has very limited use.

Green turtle

Green turtles, *Chelonia mydas*, have been called 'edible turtles' as theirs is the best and most delicately flavoured meat. Their fat is greenish in colour.

Greens are larger than hawksbills and are similarly found in warm seas worldwide. The young are carnivores but when mature they become vegetarian, grazing on algae and seagrasses in shallow waters.

The scutes on the carapace and the plastron are welded to the underlying bony layer, those on the carapaces usually displaying a beautiful, streaky pattern in colours varying from honey to greenish, and brown to black. The scutes are very thin and fragile compared with those of the hawksbill, but have been used for veneer. Green turtle shells are more likely to be polished and sold whole, still attached to the bony layer beneath, than sold as separate scutes for working.

Loggerhead turtle

Loggerhead turtles, *Caretta caretta*, are the largest of the three turtles and, like the others, are primarily found in warm seas, although they nest in more temperate areas than the other species. They have large heads and powerful jaws with which they can crush crustaceans and the other creatures that make up their prey.

The scutes on the carapace of the loggerhead are not nearly as pretty as those of the other turtles, and are as thin and fragile as those of the green turtle. Nonetheless the material has been used as tortoiseshell, though mainly as veneer. The shells are also sold whole.

Figure 8.5 Three tortoiseshell boxes: left, on white base; bottom, blond; right, natural.

Figure 8.6 Brush back with silver piqué work, and 'watermark' pattern.

Treatments and uses

Tortoiseshell is dull and slightly rough until it has been polished. The raw material displays a pattern of wavy lines on both the back and front of the scutes that are presumed to be a form of growth rings. Over a period of time the polished surface of finished tortoiseshell items dulls, and this pattern, commonly referred to as 'watermarks', reappears (Fig. 8.6). It can be removed by repolishing.

Figure 8.7 Tortoiseshell card cases: left, pressed; right, carved.

Figure 8.8 Detail of pressed card case, showing rounded edges to the pattern.

Figure 8.9 Detail of carved card case, showing sharp edges to the pattern and indications of layers in the tortoiseshell.

The following applies mostly to tortoiseshell from the hawksbill turtle, as that from the green and loggerhead is thin and more brittle, and takes less easily to the various treatments.

- Scutes from adult turtles are the most suitable for working. The shells of juveniles are often polished and sold whole.

Figure 8.10 Box combining tortoiseshell and horn. The tortoiseshell edge has flaked.

- Tortoiseshell takes a good polish, giving an almost mirror finish. This dulls with time.
- Tortoiseshell can be cut and carved. The thickest part of the scute would be used for a carving in relief, for example as the lid of a box.
- Being thermoplastic, tortoiseshell can be heated and moulded into shape, retaining this shape when it cools. Heating can either be done dry or by boiling in water. The material becomes too brittle to mould after a couple of years, so must be used fresh.
- Tortoiseshell can also be heated and pressed into a mould to produce a picture in relief (Figs 8.7 and 8.8).
- When heated sufficiently, tortoiseshell becomes slightly sticky and several layers can then be welded together. This was common practice in the past, using scrap and pieces of tortoiseshell of inferior quality to produce material that could be carved into figures, combs or boxes (Figs 8.7 and 8.9). The centre of the scutes, which were naturally thick, were not used in this way.
- Thin sections of tortoiseshell can be heated and welded to thicker pieces of horn to give more bulk at less expense. The lid and sides of a box can be made this way, giving the overall impression of a tortoiseshell item (Fig. 8.10).
- Rasping the edges of the scutes, then heating and pressing them together can produce large sheets of material. This provides pieces large enough to use as furniture veneer.
- As a veneer, tortoiseshell can be placed over a coloured or painted background. An oil-based, organic paint in white, yellow or red is

Figure 8.11 Tortoiseshell dressing table set with silver piqué work.

usually used to emphasise the colouring and pattern of the tortoiseshell (Fig. 8.5).

- Tortoiseshell can also be used as a background for inlay. The elastic quality of tortoiseshell gives it the ability to hold small pieces of gold or silver when these are pressed into the material as it is heated. It is not necessary to hollow out sections for inlay, or to glue them in place, as the material contracts on cooling. This is called 'piqué' work, and is very often seen in dressing table sets or tortoiseshell jewellery (Figs 8.6 and 8.11).
- Mother-of-pearl is also a popular material for use as an inlay in tortoiseshell, especially in items produced in the Far East.
- Tortoiseshell has frequently been used as an inlay or veneer together with ivory or brass. This is not only aesthetically pleasing, as the materials complement each other, but also protects the tortoiseshell, which can be susceptible to damage. It is seen on furniture and items such as caskets or clock cases.
- Too much heating can destroy some of the mottling and darken the material. This is sometimes apparent in small, pressed items such as boxes.
- The scutes of the green and loggerhead turtles are thin, and it would seem an obvious solution to weld them together for bulk. However, the welded material from these turtles cracks and splits with time. In Japan efforts were made to get around this by covering a single layer of tortoiseshell with plastic to add bulk, but the result was a material of inferior quality, which was therefore unsatisfactory.

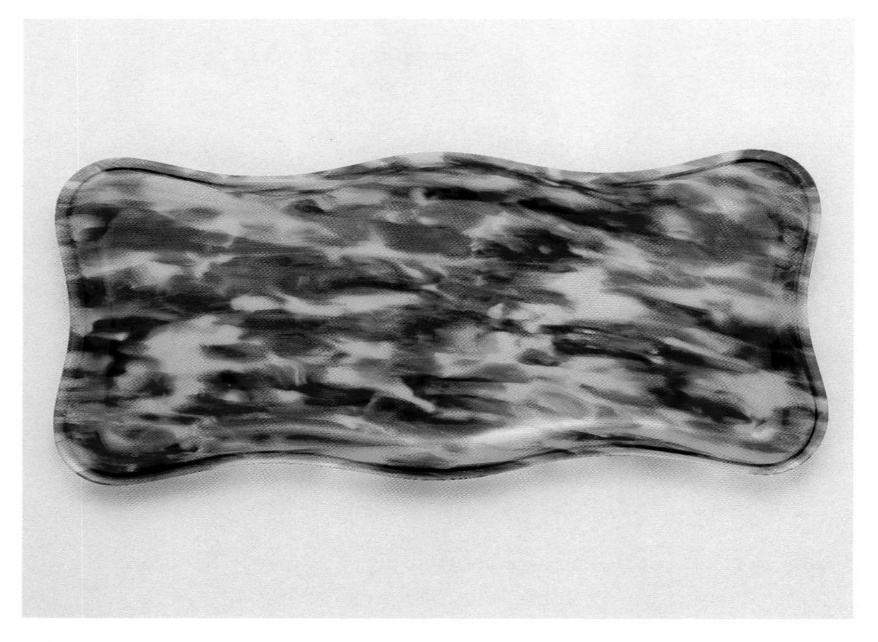

Figure 8.12 Plastic imitating tortoiseshell.

- Being thin, the tortoiseshell from green and loggerhead turtles is suitable for use as veneer. However, they are less easy than hawksbill tortoiseshell to weld together to make larger pieces.
- It is worth noting that a turtle gives not only tortoiseshell scutes, but also leather from its flippers, neck and tail, as well as meat, eggs and calipee – the cartilage used as the basic ingredient for turtle soup.

SIMULANTS

No differentiation is made here between the three types of tortoiseshell, since any imitation would automatically be of that from the hawksbill turtle.

- **Modern plastic** is the most common simulant for tortoiseshell today. Various plastics are used and made into such objects as hair ornaments – grips, clasps or combs – copying the traditional use of the material (Fig. 8.12).
- **Early plastics** made good imitations. Especially popular was 'celluloid' or cellulose nitrate. It is usually very easy to spot a plastic

Figure 8.13 Plastic imitating tortoiseshell, magnified, showing swirls of colour.

Figure 8.14 Blond tortoiseshell under crossed-polars.

Figure 8.15 Natural tortoiseshell, magnified, showing spots of pigmentation.

imitation as the colouring is not in the minute, fuzzy blobs typical of tortoiseshell, but in swirls.

- **Horn** has many of the same properties as tortoiseshell but is much cheaper and easily accessible. It has been used to add bulk to tortoiseshell by welding the two materials together, for example as a small box with tortoiseshell on the outside, or just on the lid. Horn can be dyed to resemble tortoiseshell, but this takes considerable skill. The horn is hot pressed (which partially destroys the fibrous nature of the material), then boiled in nitric acid to give it a yellow hue. After neutralising it is painted by hand. Examination shows that the colour is only on the surface, and brush strokes may be visible.

Tests and identification

Most tests on tortoiseshell are destructive, so identification is best done by sight.

Visual examination

- Viewed by transmitted light, tortoiseshell displays a pattern of irregular, fuzzy areas of brown against the honey-coloured background. They consist of minute blobs of colour, closely packed, and are continuous throughout the thickness of the material, rather than just on the surface (Figs 8.4 and 8.15).
- Plastic imitations are usually made of two colours mixed haphazardly. They are more clearly defined than the colours in tortoiseshell, and appear in swirls instead of patches (Figs 8.12 and 8.13).
- In horn the colour distribution is in blobs similar to tortoiseshell, but the pattern is different. The pale colours tend to be colder – pale beiges rather than warm, transparent honey colour. Horn also displays striations in the material due to its fibrous nature. These are absent in tortoiseshell (Figs 8.10 and 7.7).
- In finished items where the material to be examined is perhaps mounted in silver, it may not be possible to look through the material. Nonetheless, on close examination of the item with a 10× lens, it is usually possible to detect whether there are striations in the material – suggesting horn – or a lack of colour pattern in the material, which also suggests horn.
- Thin, flaky chips around the edge of the object can indicate tortoiseshell, or a tortoiseshell veneer (Fig. 8.10). Horn also flakes, but seldom in very thin layers. Plastic does not flake.
- Older pieces of tortoiseshell that have lost their high polish display so-called 'watermarks' – a faint, wavy pattern on the surface more

easily visible at an angle. This is unique to tortoiseshell. Attempts have been made to copy it in plastic by marking the plastic with little lines, but the imitation is too crude to deceive.

- Blond tortoiseshell can be difficult to identify. It lacks the striations of horn, but looks very like clear, dark honey-coloured plastic. It is usually made into fine pieces such as boxes or dressing table sets, often with gold or silver piqué work. Plastic would not be used in this way. Seen through a 10× lens by transmitted light, blond tortoiseshell looks less clear than plastic.
- The quality of the workmanship may indicate the type of material used. Tortoiseshell is unlikely to be wasted on very coarse work.
- The item for which the material has been used is also an indication of whether it is horn or tortoiseshell. Horn was seldom if ever used for dressing table sets, picture frames, card cases, and so forth. These were all traditionally made from tortoiseshell.
- Pressed and moulded objects show rounded edges, and a lack of sharp, carved edges (Fig. 8.8). Both horn and tortoiseshell can be pressed but horn will retain at least a hint of the striations typical of that material.
- Carved objects such as card cases or boxes tend to indicate tortoiseshell since other materials are too cheap to justify the work involved. There are exceptions to this rule, notably stick handles which traditionally have been carved from horn, and old powder horns. Carved edges are sharp and should show some marks from the carving tools (Fig. 8.9).
- Gold and silver piqué work is not used on plastic, and gold piqué work is unlikely to be seen on horn. Plastic may be adorned with mother-of-pearl, but this is applied with glue.
- Plastics are not pressed but are poured into a mould or extruded. This can give rise to air bubbles and marks from the joins in the mould.
- Plastics are sometimes moulded in such a way that the back of the item, when viewed by transmitted light, seems to give a fuzzy blob effect in the colours. This surface effect shows up when viewed by reflected light.
- Obvious though it may seem, it would be unlikely to see 'Made in England' stamped on the base of a piece of tortoiseshell, but this, or similar, is often present on plastic imitations.

Tests

- **Ultraviolet light.** The first of the two tests that does not in any way harm the material is to view it under UV light. Both tortoiseshell and horn show a slight fluorescence, especially in the blond areas, whereas plastic – with the exception of casein – is inert.
- **Crossed-polars.** The second harmless test is to view the material

under crossed polars. With the aid of magnification, horn and tortoiseshell show lines and spots in all the colours of the spectrum. Plastic can be inert, but it can also display multicoloured swirls caused by the stress in the material. This test is especially good for blond tortoiseshell (Fig. 8.14). However, it is dependent upon being physically able to get the material to be tested onto a polariscope. Finished items may be too large, and mounted items will not allow light to pass through them.

- **Sectility.** This is not a useful test as tortoiseshell, horn and plastics can all be pared with a sharp knife
- **Burning and hot-point tests.** Tortoiseshell burns at a lower temperature than horn. A sliver of either will burn with the typical small of burning hair because they are basically made of the same material. It is a destructive test and so it is preferable to cut a minute sliver of the material from an area where it will not be noticed, rather than stick a hot needle into the item. Plastics give off a typical, acrid smell when they burn.

Note: A further reason for burning only a small sliver of the material is that the most common early plastic used as an imitation of tortoiseshell was celluloid (cellulose nitrate), *a very highly inflammable product.* This material catches light with an explosion and immense care must be taken if there is any possibility whatsoever that the item to be tested could be made of celluloid. Also, some plastics melt quickly so a hot needle can cause a lot of damage to an object.

CONSERVATION STATUS AND AVAILABILITY

Marine turtles have been around since the earliest dinosaurs, but having survived for 200 000 years they are now under threat of extinction. Their natural predators are many and varied. Some predators eat the turtle eggs before they are hatched, some eat the tiny hatchlings, and there are plenty more creatures that will eat a young turtle in the sea. But their worst enemies are humans.

For centuries humans have caught turtles to eat. Their meat and eggs were valuable sources of protein, and the skin from their necks and flippers can be turned into leather. However, when tortoiseshell became popular as a decorative material, the hunting became an industry, and the hawksbill turtles were fished, or caught on the beaches as they tried to lay their eggs. Tens of thousands were caught every year. Greens were also caught in their thousands, but this was for the calipee (the cartilage around the bony interior layer of the plastron), which was used to make the famed turtle soup.

Turtles are susceptible to pollution from, for example, oil rigs and shipping. They get caught in fishing nets, and their nesting beaches are destroyed. Tourism has brought many hazards for the turtles. New hotels encroach on the beaches, and the lights from the hotels or nearby roads disorientate the tiny hatchlings, so that they cannot find their way to the sea. A turtle does not mature and lay eggs until it is about 20 to 30 years old. Although it is impossible to put a definite figure on their survival rate, it is reckoned that, in the case of the hawksbill turtle, only 1 in 8000 hatchlings survives to adulthood. It will take many years for severely depleted turtle stocks to be built up again.

Turtle farming has been attempted in the Cayman Islands, but with little success for the hawksbill. There has been much more success with green turtles, but although the shell is thicker than that of wild green turtles, the material is not of very good quality and it tends to crack. It also has a high salt content, which makes it unsuitable for moulding.

Today some countries still allow fishing of hawksbill turtles for their shell, though thanks to closed seasons and other types of management, the amount fished is greatly reduced. Most of the CITES signatories respect the total ban on the international trade of marine turtles or their parts. Thus, although it may be possible to purchase items made from tortoiseshell in, for example, the Far East, it is not permissible to import them into Europe, America, and many other countries.

Past and present uses

Marine turtles have primarily been a source of protein: eggs and meat. To local populations in the areas where turtles lived these were part of their everyday diet, and the skin of the flippers, neck and tail also made excellent leather.

Turtles were a highly prized catch for some centuries on sailing vessels, as the animals could live for a long time without food or water. They were carried for long periods on the ships before finally being slaughtered and eaten, providing that rare commodity: fresh meat.

In many countries the turtle has symbolised fertility and immortality. In Hawaii and Burma they have been worshipped as deities, while in China their longevity (and apparent indestructibility) was much admired. It is said that the foundations of some palaces and temples in the Forbidden City rest on turtles, buried there alive to protect the buildings.

The scutes of all types of terrapins, turtles and tortoises have been tried out at one time or another as a form of tortoiseshell, but without success. Unfortunately not much was known about the different species, and, as some of them were able to regrow scutes that had fallen off, it was presumed that hawksbill turtles would also be able to do so.

The scutes are removed from the bony shell by heat, so it was a common practice to catch a turtle and hold it over a fire till the scutes fell off, then return the animal to the sea. Unable to regrow its protective layer of keratin, it died.

Although there are no illustrations in the tombs of ancient Egypt of tortoiseshell in use, there are pictures of marine turtles, and it is thought that the material was used for decorative purposes, for example knife handles or bracelets. Like horn, tortoiseshell does not survive the passage of time.

In Europe the Romans used it as a veneer on furniture, door frames, babies' cradles and even on baths, but, being rare and expensive, it was not much seen again until the end of the seventeenth century, when it found popularity with the royal courts for use in marquetry. The greatest master of the craft was André Charles Boulle, who perfected the art of combining tortoiseshell with brass and wood – usually ebony – in intricate designs. He produced desks, cabinets, tables and all manner of furniture, now commonly referred to as 'Boulle furniture'. Born into a family of cabinet-makers, he was appointed to the royal court at a young age. While only in his thirties he produced large amounts of furniture for the palace at Versailles, where examples can still be seen.

Another use for tortoiseshell veneer around this time was for clock cases, both long-case and bracket clocks. Later its use became more widespread, if somewhat less impressive, and boxes and caskets with tortoiseshell veneer and ivory edges, corners and keyholes could be found in many wealthy homes. Tea-caddies in this style were particularly popular.

It would seem that while in the west tortoiseshell was made into anything from trinkets to large pieces of furniture, in the east the shell was used more delicately. In museums in the Far East can be seen boxes, trays, small tables and even wooden slippers which were made with tortoiseshell veneer, intricately inlaid with mother-of-pearl, and which depict landscapes or flowers in a style typical of the region.

The Imperial Collection in Japan houses items of tortoiseshell that are over 1000 years old, but these are probably of Chinese origin. The Japanese tortoiseshell industry started only around 300 years ago. Much of what has been produced in Japan has been for export as their craftsmanship was unparalleled and their work much in demand in the west. The Japanese term for tortoiseshell is 'bekko'.

A completely different use was found for tortoiseshell in the islands between New Guinea and Australia. There it was a popular material from which to make hunting or ceremonial masks, combined with other materials such as sea shells and feathers. The masks often represented animals, for example birds or fishes (Fig. 8.16).

One of the most common uses for tortoiseshell has always been hair combs. Whether of the type for personal grooming, or for hair adorn-

Figure 8.16 Tortoiseshell mask. Mabuiag Island, Torres Straits. Eighteenth century. *The British Museum, London.*

ment, they were used in many countries worldwide. In Spain they were used to hold the mantilla in place, in Japan to hold elaborate hair styles in place. Some combs were plain, while some were carved or inlaid.

Tortoiseshell as an inlay was popular for adorning musical instruments, especially string instruments, where it was often combined with ivory. It was also used as decoration on the butts of guns and rifles, where again it was often combined with other materials such as ivory or mother-of-pearl.

Though still rare and expensive, tortoiseshell was a favourite material in Europe by the twentieth century. It was made into dressing table sets, mounted in silver and inlaid with piqué work. Card cases were sometimes made of the pressed material, and sometimes finely carved. Cane and umbrella handles, small boxes, spectacle frames and cases, fans, opera glasses, and many other similar objects could be made from, or adorned with, this attractive material.

Many of these items come onto the market via antique shops or auction houses today, often in good condition. Tortoiseshell items that have lost their lustre and dulled with age can be professionally re-polished.

9 Pearl

A pearl is a concretion produced by a mollusc. There are a number of theories as to why a pearl forms, but it seems evident that it is, at least in part, as a reaction to some type of intruder or foreign body in the shell.

There is a huge variety of pearls. Some are uneven in shape with a dull surface, others are coloured. The majority are nacreous, that is to say they are covered with mother-of-pearl and have the pearly lustre usually associated with this gem.

All pearls can occur naturally, but today the majority of them are cultivated and are known as 'cultured pearls'. Although they grow naturally inside a mollusc, their growth is instigated and controlled by human intervention.

Structure and properties

Pearls with a nacreous surface can be divided into three groups: natural pearls, nucleated cultured pearls, and non-nucleated cultured pearls. (Pearls without nacre will be dealt with separately at the end of the 'Species' section.) A further group – blister pearls – are hemispherical pearls that grow attached to the oyster's shell. These can also be cultured.

It is also possible to divide the groups into marine and freshwater pearls. The molluscs that produce these two types of pearls are termed marine pearl oysters and freshwater mussels.

To understand the structure of a pearl, it is first necessary to understand the structure of the mollusc from which it comes.

The vast majority of pearl producing molluscs belong to the group called bivalves (see also Chapter 10, 'Shell'), which have two shells, hinged together, that can open and close as the animal feeds. Inside the shell is the soft body of the mollusc. It has no head, and a very simple nervous system. The main part of the body is called the mantle, which covers all the organs, and which secretes the shell.

The shells are made up of a slightly horny covering called the periostracum, and two or three layers of calcium carbonate. These layers can crystallise in different ways, according to the species: columnar, prismatic or platelet form, or a combination of these. The crystals form in a framework of organic matrix.

In pearl-bearing molluscs the inner layer is nacreous, and is called 'mother-of-pearl'. It is composed of minute, overlapping platelets of calcium carbonate in the form of aragonite, which allow incidental light to be diffracted, giving the layer its unique appearance (Fig. 10.7).

It is this ability to secrete the organic and inorganic compounds that make up the shell, which gives the mollusc the ability to secrete the same substances to produce pearls. All pearls consist of approximately 90 per cent calcium carbonate, plus water, and an organic matrix. As with the nacreous inner layer of the shell, the calcium carbonate in pearls forms as aragonite platelets and gives the pearl its nacreous lustre (sometimes called 'orient').

Nacreous pearls come in a large variety of colours, from white to black. At the white end of the scale are the creams, champagnes, and pinks. At the other end are the greys and blacks, with iridescent tinges of blue, purple, green and pink. All these colours can occur naturally. The colour of the black pearls is due to the presence of the organic pigment porphyrin. These are only produced by certain oysters. The colours of other pearls are also basically dependent upon the oysters that produce them and on the method used to cultivate them, though other elements, such as the minerals in the water where they are growing, may have some effect.

Natural pearls form by chance within the oyster, and are not induced by humans. They consist of sequential concentric layers of nacre, with a resulting structure similar to that of an onion. This contrasts with shell nacre (mother-of-pearl), which is laid down in parallel layers.

Figure 9.1 Various nucleated cultured pearls.

Natural pearls probably occur when a form of foreign body enters the shell, and the mollusc is unable to get rid of it. It was originally believed that the irritant was a grain of sand, but this is now considered extremely unlikely, and the actual cause is open to conjecture. Oysters feed by taking in water and filtering out the food, and are adept at filtering out sand and other unwanted particles, so there is a possibility that the foreign body is a form of parasite. When the parasite dies and decays, the mollusc encapsulates it in nacre, reducing the irritation to the soft part of its own body.

Nucleated cultured pearls, as the name implies, consist of solid nuclei (usually mother-of-pearl shell beads) with an overlying 'skin' of nacre (Fig. 9.1). As with shell nacre, the organic matrix secreted by the mollusc forms a framework over the bead, in which the aragonite crystallises. This organic material seems to be concentrated at certain growth stages and also to be influenced by seasonal change. This fact aids the identification of the pearls. For example, it is the cause of the demarcation seen between the nacre and the bead in a nucleated cultured pearl.

Non-nucleated cultured pearls have a similar structure to that of natural pearls, but at their core there is an irregular cavity, which is a result of the initial stages of their growth which was induced by humans.

Natural blister pearls can occur when a parasite bores through the mollusc's shell. The mollusc seeks to protect itself by increasing the rate of nacre secretion at the point of invasion. This leads to a bump on the inside wall of the shell, which may finally become a hemispherical blister pearl, or which may simply outline the shape of the parasite (Fig. 9.2).

Composite cultured blister pearls are composed of a hemisphere (or other shape) of nacre that has been cut out from the shell, where it had formed to cover a nucleus that had been inserted artificially and glued to the shell. It is filled with resin or wax, and backed with a slice of mother-of-pearl. The hemispherical blister pearls are often called 'mabés' (Figs 9.2 and 9.3).

Species

The taxonomy of pearls can be a little uncertain and the borders between the families a little blurred. Some of the following are occasionally given other Latin names than those quoted here.

Marine pearl oysters

The **white-, silver-** or **golden-lipped pearl oyster**, *Pinctada maxima,* is the largest pearl producing oyster, and can measure up to 30

Figure 9.2 Broken blister pearl in shell, showing nucleus, and parasite causing natural blister in shell.

Figure 9.3 Four composite nucleated cultured blister pearls ('mabés'), colour enhanced, and a pair of natural coloured mabés on the right.

Figure 9.4 South Sea nucleated cultured pearl (magnified).

centimetres across. The nacre inside the shell is silvery in colour, and in the case of the golden-lipped oyster, it is tinged with a creamy gold colour around the edge. It lives in the East Indian Ocean and the west Pacific, and is the oyster used in the pearl farming area of north-western Australia. It is used to produce South Sea pearls, which are nucleated cultured pearls, measuring 10 to 20 millimetres in diameter, and are white to golden in colour (Fig. 9.4).

The **black-lipped pearl oyster**, *Pinctada margaritifera,* is a little smaller, and measures from 15 to 25 centimetres across. It, too, has silvery nacre inside the shell, becoming silvery grey at the edges, which are surrounded on the lip of the shell by a black border that resembles enamel. The oyster lives in the Indian Ocean and the western and central Pacific, especially around French Polynesia. It is used to produce the famous **Tahitian black pearls**, which are nucleated cultured pearls. The nacre secreted by these pearl oysters is naturally coloured and varies from light grey to black, with iridescent hints of green, purple, pink and blue. The pearls range in size from 10 to 18 millimetres in diameter.

The **Akoya pearl oyster**, *Pinctada fucata martensii,* is again smaller, measuring 8 to 10 centimetres across. It has silvery nacre inside the shell. Akoyas are used in Japan to produce their world-famous nucleated cultured pearls, and are now also being used in China and other parts of the Far East. The cultured pearls produced by these oysters vary in size from 2 to 10 millimetres in diameter, and are white to cream in colour.

The **black-winged pearl oyster**, *Pteria penguin,* has silvery nacre, a wide black lip and a narrow 'wing' shape where the shell hinges. It measures from 10 to 25 centimetres across the main body of the shell. The colouring and shape give the species its Latin name. It lives in the Red Sea, the Indian Ocean and the western Pacific. It is this oyster that produced the original **mabé pearls** before the word became generic, and it is still used to produce mabé cultured pearls. These cultured composite blister pearls are white, and measure from 20 to 25 millimetres across.

The above are the most common marine pearl oysters that are farmed today. There are many more varieties, not all of which are cultivated. The following are of historic interest.

The **La Paz pearl oyster**, *Pinctada mazatlanica,* lives in the eastern Pacific, especially around Central America. It gives white to grey pearls and was probably the source of the famous 'La Peregrina' natural pearl (see 'Past and present uses' section).

The **Atlantic pearl oyster**, *Pinctada imbricata,* at between 5 and 7 centimetres across, is a relatively small oyster. It was the source of the first natural pearls that were traded from Venezuela. It is not farmed.

The **Ceylon pearl oyster**, *Pinctada radiata,* is historically an important oyster as it is the species that gave the natural pearls that were fished around the Arabian Gulf, the Red Sea, and the Indian Ocean.

Freshwater pearl mussels

The **Biwa pearl mussel**, *Hyriopsis schlegelii,* is probably the most famous of the freshwater mussels. It was native to Lake Biwa in Japan, though the stocks are now very depleted. It was used in the production of **Biwa pearls**, which were the first freshwater, non-nucleated cultured pearls to come on the market. They were of an irregular oval shape.

The **cockscomb pearl mussel**, *Cristaria plicata,* lives in Japan and China. It produced the original 'rice-krispie' pearls. They were small, irregular in shape, and had a wrinkled appearance. They were naturally white, but were frequently dyed. The mussels have thin shells and, when pearl cultivation was first attempted in China, were proved to be incapable of producing freshwater cultured pearls of high enough quality.

The **triangleshell pearl mussel**, *Hyriopsis cumingii,* is the oyster now used in China to produce their freshwater cultured pearls, which are of a vastly superior quality to their original 'rice-krispies'. The triangleshell is also farmed in Japan. It has a much thicker shell than the cockscomb, and it can produce nearly perfect round pearls with a good lustre. The naturally coloured nacre occurs in whites, orange-browns and blue-greys. White is the most popular colour (Fig. 9.5).

The **European pearl oyster**, *Margaritifera margaritifera,* is of historic interest as it was the source of most of the European freshwater natural pearls. It was native to Europe and parts of North

Figure 9.5 Chinese freshwater pearls.

America. The pearls were plentiful, and were smooth and silvery-white, with a good lustre. It is now almost eradicated from European rivers due to overfishing.

Unusual pearl types

The term **'baroque pearl'** is used to describe an irregularly shaped pearl. It was first applied to irregular natural pearls, but can also be used to describe the irregular shapes seen in cultured pearls. Baroques can be very beautiful and are highly prized (Fig. 9.6).

Figure 9.6 Baroque pearl in jewellery.

Figure 9.7 Unusual cultured pearl shapes.

There are many natural and cultured pearls of interesting shapes that come onto the market. The freshwater mussels of the eastern United States produce natural pearls that are usually silvery white, often have a slightly silky sheen rather than a pearly lustre, and can be an amazing variety of shapes (Fig. 9.7).

Circlée pearls are nucleated cultured pearls from marine pearl oysters. The round to oval pearls display indented rings around the middle of the pearl. It is not known exactly why this patterning occurs. Two possible theories are that the mantle tissue was twisted when inserted into the oyster, or that the pearl turned frequently while it was growing.

The term **keshi pearls** was first applied to tiny pearls that were recovered from Akoya oysters along with the harvest of nucleated cultured pearls. Although they are evidently induced by the culturing process, they occur by accident and are a by-product of it. They are smaller than the nucleated pearls and develop elsewhere inside the shell, at the same time as the nucleated cultured pearl. It is not known why they develop, but they could be the result of a piece of the mantle tissue breaking away from the graft, and setting up an irritation elsewhere. Today the name keshi – which means 'poppy seed' in Japanese – refers mainly to non-nucleated pearls occurring in the shells of South Sea or Tahitian oysters.

Seed pearls are natural pearls that are less than 2 millimetres in diameter. They are a white or creamy colour, and were very popular in the nineteenth century. They were usually imported from India (Fig. 9.23).

Abalone pearls from the abalone, of the genus *Haliotis*, are unusual in that they come from the group of molluscs called gastropods, and not from bivalves. The molluscs have only one shell (see Chapter 10,

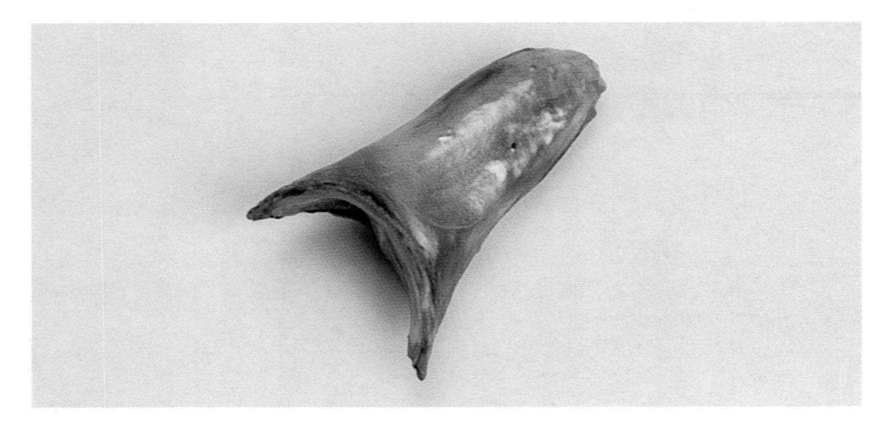

Figure 9.8 Abalone pearl (magnified).

'Shell'). The pearls are iridescent, with the same blues, greens, purples and pinks that are displayed by the shells. The pearls are natural, and are usually hollow and of unusual shape – often cusp shaped – and very seldom round (Fig. 9.8). Nowadays cultured composite blister pearls are being produced in abalones.

Non-nacreous pearls

There are several other molluscs that can, on occasion, also produce pearls. The most famous are the conch and the baler, which both produce pearls of great value, although they are non-nacreous. Others, such as the horse conch and the noble pen shell, are less well known. Some molluscs produce concretions that are called pearls, but which tend to be curiosities rather than beautiful items. The most famous is the clam.

The **conch**, *Strombus gigas*, is a gastropod. The lining of a young shell is a peachy-orange colour, becoming strong pink as the mollusc reaches maturity. When viewed under magnification the surface displays a flame-like pattern, caused by the striated appearance of the aragonite crystals that make up the inner layer of the shell. This pattern is repeated on the pearls (Fig. 9.9).

Conch pearls are very rare, slightly oval in shape, and are seldom more than 8 millimetres long. They have the same orange or pink

Figure 9.9 Conch pearl, showing 'flame' pattern (magnified).

colour as the shell, and, like the shell, can fade to pure white if exposed to too much light. For this reason they are best stored in the dark.

The **baler**, *Melo melo*, and other molluscs of the same genus, are also gastropods and can also produce pearls. The baler shell is orange inside and has the same flame pattern as a conch shell. The pearls can vary in size, some being quite large. They display the same colouring and flame pattern as the shell.

The **clam**, a bivalve of the genus *Tridacna*, occasionally produces pearls. They are uneven in shape and seldom show even a hint of pearly lustre. A large clam can develop a very large pearl – the largest known is 23 centimetres long and weighs over 6 kilograms – but it is not so much a thing of beauty, as a very interesting specimen.

Pearl cultivation

The first cultivation process was a system developed by the Japanese company Mikimoto, and it is still the method currently used for culturing Akoya pearls.

To produce these, mature oysters are taken from the water and dipped into a mild anaesthetic to make them gape. Each oyster is put

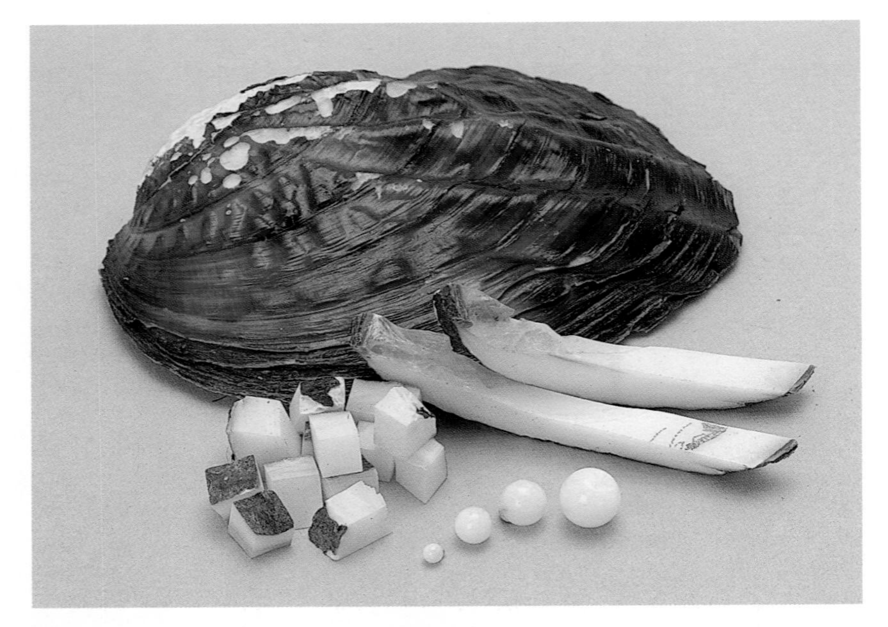

Figure 9.10 Washboard mussel shell, sliced and cubed for making bead nuclei.

in a clamp, wedged open, and two small slits are made in the gonad. A little strip of mantle from a sacrificed oyster is inserted into each of these slits, followed by a bead nucleus fashioned from the shell of a freshwater mussel (traditionally the Mississippi Pigtoe or Washboard) (Fig. 9.10). The oyster is put back into the water – an unpolluted area such as a lagoon – where the inserted mantle grafts form 'pearl sacs' which encapsulate the nuclei and produce the nacre to cover the beads. This takes from six months to two years, depending on the quality of cultured pearl required. The thicker the nacre, the better the lustre of the pearl and the more beautiful its appearance.

The thickness of a pearl's nacre depends not only on the length of time the oyster is left in the water, but also upon the health of the host oyster, the surrounding water conditions, and the quality of the bead nucleus used. The season at which the cultured pearls are harvested also has an influence on the lustre of the nacre.

South Sea pearls are large, and are grown in, for example, the silver- and golden-lipped oysters, which are much larger than the Akoyas. South Sea pearls are also nucleated with mother-of-pearl beads, but only one is inserted at a time. They produce nacre faster than Akoyas and are left in the sea for two years, resulting in the thicker nacreous layer that is typical of South Sea pearls (Fig. 9.4). The oysters are suspended from off-shore floats, at a depth of 10 to 15 metres, and are tended and harvested by divers who sail out to the oyster beds on boats equipped as laboratories.

Many years after the original pearl cultivation was started, it was realised that it was not necessary to insert a hard nucleus into an oyster to induce it to produce a pearl, and that it was sufficient simply to insert a piece of mantle. This is the method favoured by the Chinese, and gives 'non-nucleated cultured pearls'. The pearls are grown in freshwater mussels, suspended from floats in lakes and rivers. Up to 30 mantle grafts are inserted into a single host mussel's mantle tissue, just inside the shell.

The original attempts at Chinese pearl cultivation produced pearls which were relatively small and wrinkled. They were generally referred to as 'rice krispies' after a well-known breakfast cereal of a similar shape. Nowadays much better quality non-nucleated cultured pearls are produced in China (Fig. 9.5).

With all types of cultured pearl, the host oysters and mussels must be tended during the cultivating process. They are regularly lifted from the water and cleaned to remove algae and barnacles that may have accumulated. When the appropriate time has passed for the pearls to have reached their optimum quality, they are harvested. The oysters or mussels are opened and the pearls are extracted.

Not all the molluscs will have produced pearls, and among those that have, many pearls will be of poor quality. It is an uncertain science

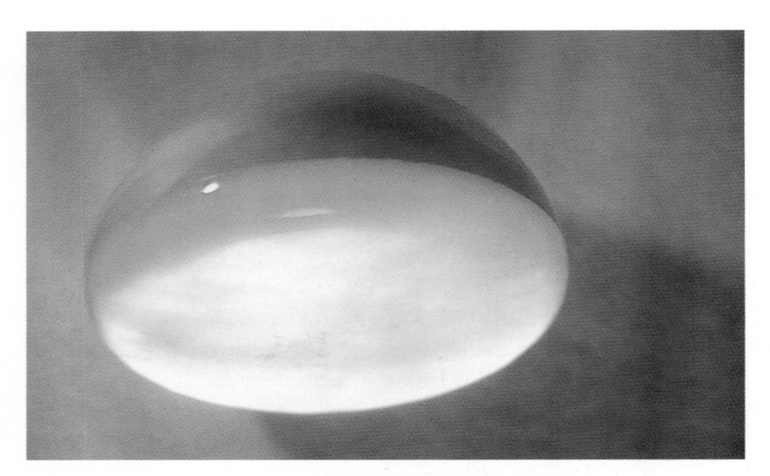

Figure 9.11 Composite nucleated cultured blister pearl ('mabé'), showing the mother-of-pearl back (magnified).

and the success rate is not high. For example, Akoyas will produce about 200 saleable pearls and 50 top-quality ones from 1000 grafted oysters.

Some species of oyster, for example the South Sea pearl oyster, can be nucleated a second time, immediately after the first pearl has been removed, by inserting a new bead into the pearl sac already there. Akoyas can only be nucleated once.

Marine pearl oysters that are farmed are all nucleated with a bead and a piece of mantle tissue. Freshwater mussels that are farmed are implanted with mantle tissue grafts only. In some areas the oysters and mussels to be used are bred and cultivated until they reach maturity and can be implanted. In other areas oysters are caught in the wild to be cultivated.

As already stated, a mollusc can also cover an object that is attached to the inside of its shell. This can be a natural occurrence, but is also copied to produce cultured blister pearls. A hemisphere of mother-of-pearl or plastic is attached to the shell underneath the mantle, and the mollusc is left to coat the object with nacre. Usually the blister pearl will be cut away from the shell and the shell back and nucleus removed, leaving a hollow hemisphere of nacre. This is cleaned, and filled with resin or wax. The hemisphere is then backed with a new piece of polished mother-of-pearl. The result is a composite cultured blister pearl. The term 'mabé' is now generic and used to describe these pearls, although, strictly speaking, it should only refer to blister pearls from a single species of oyster – the black winged oyster (Figs 9.3 and 9.11).

Figure 9.12 Dyed Chinese freshwater pearls.

TREATMENTS AND USES

A good pearl comes out of the oyster complete and ready for use. It needs no treatment other than washing, and having a hole drilled for stringing or mounting purposes. This makes it unique among gem and ornamental materials.

- In times past, pearls were simply washed in water and buffed before use. Today most of them are washed in very dilute hydrogen peroxide, which is said to bring out their natural colour.

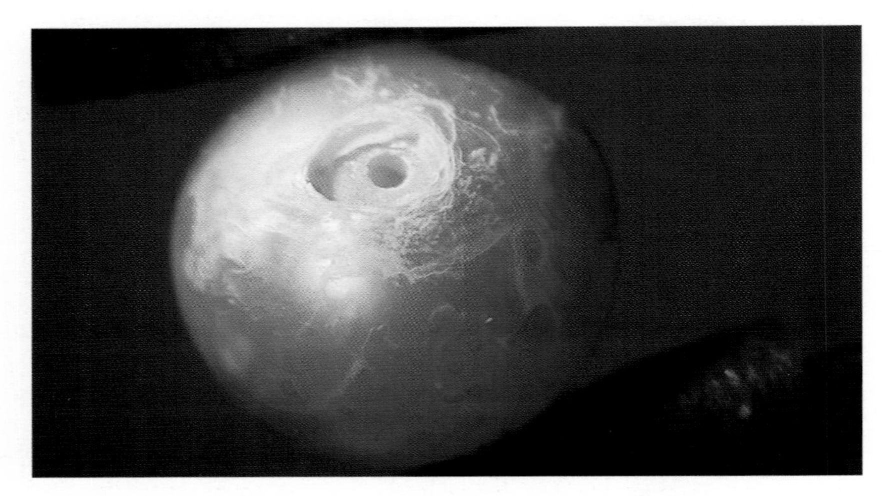

Figure 9.13 Nucleated cultured pearl destroyed by bleaching.

Figure 9.14 Dyed black nucleated cultured pearls.

- Pearls can be soaked in hydrogen peroxide to bleach them. This used to be routine with Chinese freshwater pearls as their colour was not good, but is probably less necessary today as the quality of the pearls from China is much improved. Bleaching tended to dry out the organic matrix in the pearls and cause them to become brittle (Fig. 9.13).
- Pearls can be dyed. Nowadays they can be dyed any colour, including totally unconvincing ones, such as bright blue. More usually,

Figure 9.15 Nucleated cultured pearl with broken nacre.

the dye colours chosen are the ones that can occur naturally, such as pale pink tints (Fig. 9.12).

- Black is a popular colour to dye pearls, imitating Tahitian black pearls. This used to be done by staining the pearls in silver nitrate, which turned black when exposed to the light. Today it is more normal to use aniline dyes (Fig. 9.14).
- It is also possible to turn cultured pearls grey by irradiating them. This darkens the mother-of-pearl bead nucleus in nucleated cultured pearls, but not the nacre produced by the marine oyster.
- Freshwater cultured pearls can also be turned grey by irradiation.
- White composite cultured blister pearls (mabés) can be painted on the concave inner surface before it is plugged with wax or resin and backed with mother-of-pearl. The added colour is almost impossible to detect and simply gives a warmer tint to the nacre (Fig. 9.3). They can also be painted with an imitation pearl coating to give the impression of a deeper lustre.
- White composite cultured blister pearls can also be dyed (Fig. 9.3).
- Pearls are drilled right through the centre for stringing as beads, or part way through to be mounted and glued on a pin, in, for example, a ring mount.
- A new innovation is faceting pearls. It can only be done to nucleated pearls with a thick enough layer of nacre, or to non-nucleated pearls which are solid nacre. The surface of the pearl is polished into masses of flat surfaces, resulting in a pearl that looks more like a polished, mineral gemstone, but with a pearly lustre.
- Pearls are only durable given the right conditions. Chlorine in water will dry them out and cause them to crack (Fig. 9.13). Liquids such as hairspray will also destroy them.
- Pearls are relatively soft, and the nacre on cultured, nucleated pearls can wear thin with use.

- Nucleated pearls are more susceptible to breaking than those composed entirely of nacre, and a sharp blow can crack the nacre, exposing the mother-of-pearl nucleus beneath (Fig. 9.15).

SIMULANTS

- **Shell beads** are made from the same material as the pearls themselves, so are naturally nacreous. However, the nacre does not cover the entire bead, but appears in layers due to the structure of the shell (Fig. 10.13). When dyed black, shell beads make convincing imitations of black pearls, though they tend to be too regular in shape.

Figure 9.16 Plastic beads coated with mother-of-pearl simulant.

- **Essence d'orient.** Possibly the most convincing of the imitation pearls were those using this seventeenth century invention. Essence d'orient was an iridescent liquid, which was made by grinding the scales of the bleak fish, and suspending them in lacquer. The mixture was then painted onto the inside of beads of thin glass, using several coats. The beads were filled with wax to give them a more convincing weight. An easier method of making the beads was to paint the essence d'orient on the outside, but it was far less convincing (Fig. 9.16). Today the liquid is still used, but herring scales are used instead of bleak fish, and a synthetic resin is used instead of lacquer. The nucleus can still be of glass, though mother-of-pearl or plastic beads are also used.
- **Mica** can be used today instead of essence d'orient. Tiny flakes of the material are suspended in a synthetic resin and applied in the same way.
- **Other compound coatings.** A number of other compounds can be induced to crystallise in a submicroscopic platelet form imitating aragonite, and suspended in a variety of lacquers or resins, which can be applied in the same way as essence d'orient.
- **Plastics** appear as a simulant for every organic gem material. They have also been used for imitating pearls, but, unless coated in the ways mentioned above, both early and modern plastics are totally unconvincing.

Note: Non-nucleated cultured pearls of poor quality are nowadays plentiful and cheap. It is likely that they will replace imitation pearls on the market as there will be no economic advantages in producing pearl simulants.

TESTS AND IDENTIFICATION

Pearls are very susceptible to damage from chemicals, and testing is therefore limited. An expert who is used to dealing with pearls will have a good idea of what he is looking at, simply by handling the pearls. For the less experienced, more careful examination is necessary. Many pearls can be identified by sight, but in the past few years this has proved more and more difficult. It is often possible to see if a pearl is dyed, or if it is a nucleated cultured pearl. Non-nucleated cultured pearls present much bigger problems as they have the same structure as natural pearls, and are now available in much better qualities than was previously the case.

Figure 9.17 Cultured pearl, showing bead nucleus.

Figure 9.18 Nucleated cultured pearl showing uneven layer of organic matrix.

Visual examination

- A collection of pearls, for example as a necklace, which is perfectly matched in colour, is more typical of cultured pearls. Natural pearls vary in colour and would not be bleached or dyed to correct this.
- A collection of pearls that are perfectly matched in size would similarly indicate cultured pearls.

Figure 9.19 Dyed nucleated cultured pearl, showing accumulation of dye.

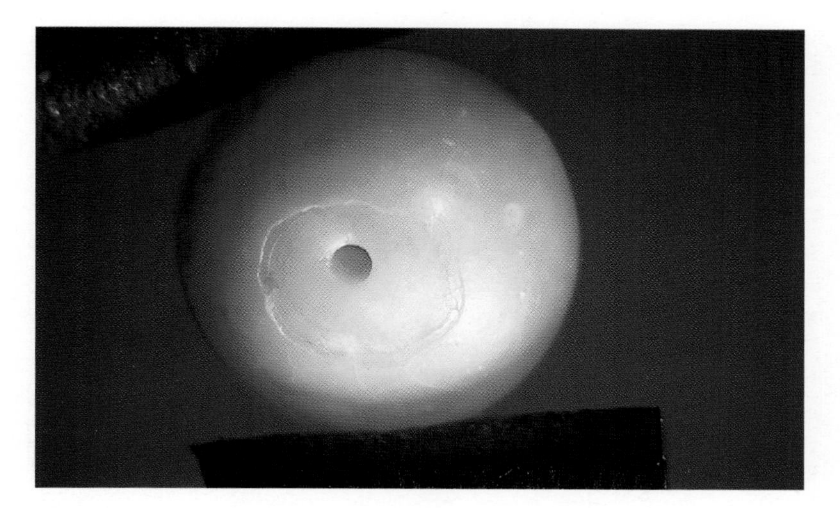

Figure 9.20 Nucleated cultured pearl with chipped nacre.

- A collection of circlée pearls would suggest nucleated cultured pearls. Circlées are rare as natural pearls.
- Viewed by transmitted light and turned in front of it, nucleated cultured pearls may display what is known as 'candling'. The light shines unevenly through the pearls and it may be possible to see lighter and darker areas running across each pearl. This effect is due to the mother-of-pearl bead used as a nucleus, which has crystals running in straight layers, and not in radiating layers. It is easier to see the effect

on pearls where the nacre is thin, either due to a thin layer being deposited during cultivation, or where it has worn thin through use.

- Viewed with a 10× lens the drill hole of a nucleated cultured pearl should reveal the presence of the mother-of-pearl bead nucleus. The demarcation between the bead and the overlying nacre should be evident. (If the bead is fully drilled, this demarcation should show at both ends of the drill hole) (Fig. 9.17).
- If a greater than normal amount of the dark organic matrix was deposited around the bead nucleus in a nucleated cultured pearl (before the oyster started secreting nacre), the demarcation line between the layers will be more marked, and a dark line or layer will be visible down the drill hole. This thick layer can also give a greyish appearance to the pearl. Further, the organic matrix tends to be deposited haphazardly, which leads to the perceived greyish colour being patchy (Fig. 9.18).
- A nucleus of a different colour to the nacre is an obvious sign of a nucleated cultured pearl. A dark bead beneath white nacre will give the impression of a grey pearl. It is probable that the pearl has been irradiated as this process darkens the freshwater mother-of-pearl bead rather than the nacre produced by a marine oyster. It is also possible that the bead has been dyed.
- A dyed pearl, whether natural or cultured, may show concentrations of colour in surface imperfections, or between nacre layers when viewed down the drill hole. This may be especially evident in nucleated cultured pearls where the dye can accumulate between the mother-of-pearl bead and the overlying nacre (Fig. 9.19).
- A dyed composite blister pearl may show accumulations of dye around the join of the top section and the mother-of-pearl back. Those that have been colour enhanced by painting the inside of the nacreous top are almost impossible to detect.
- As a result of wear and tear, the nacre on a nucleated cultured pearl can occasionally become chipped. This is most likely to be apparent around the drill hole (Fig. 9.20).
- Signs of flaking in thin 'nacre', especially around the drill hole, may indicate a coated pearl simulant (Figs 9.21 and 9.22).
- A slightly glassy lustre may indicate a hollow glass bead with an imitation pearl coating applied to the inner surface. This must be examined carefully under magnification, where it should be possible to see both surfaces of the glass.
- Conch and baler pearls display a flame-like pattern when viewed under magnification (Fig. 9.9). This pattern appears over the surface of the whole pearl, from every angle. The pattern would be missing from some areas in items fashioned from the conch or baler shell itself. Instead there would be evidence of the banded structure of the carved shell (see Chapter 10, 'Shell').

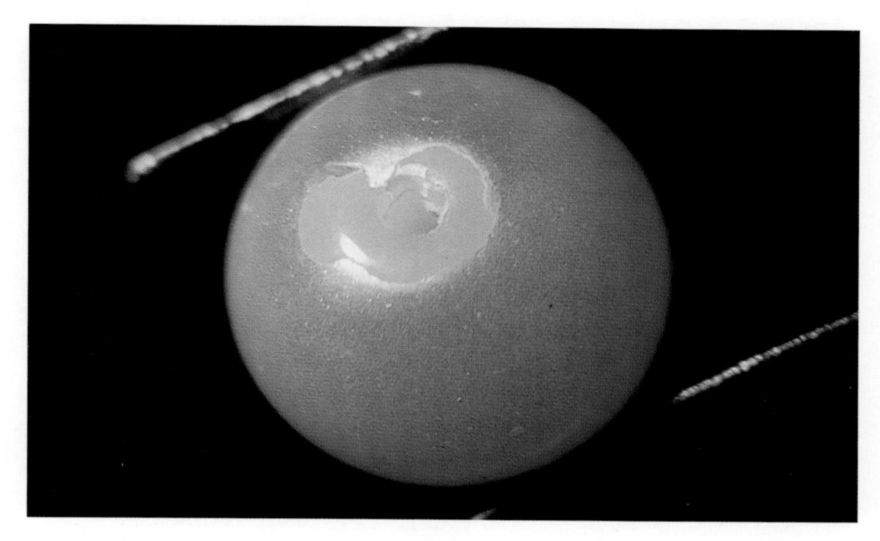

Figure 9.21 Coated plastic bead showing surface wear around drill hole.

Figure 9.22 Coated plastic bead with worn surface.

Tests

- **Ultraviolet light.** Pearls from marine oysters fluoresce a chalky-white under UV light, while some freshwater pearls fluoresce a yellowish colour. However, coated plastic pearls can also fluoresce a yellow colour. As shell – and therefore shell beads – have the same fluorescence as marine pearls, the test is not conclusive.

There is an old test for pearls that involves rubbing them gently across the teeth. It is not a destructive test, but it is unhygienic.

- **Teeth test.** Pearls that feel slightly gritty when rubbed across the teeth are probably either cultured or natural. The gritty effect is caused by the minute overlapping platelets of aragonite that make up the pearls' nacre. Pearls that feel completely smooth when rubbed across the teeth are probably simulants.

For further tests it is necessary to seek the help of a gem-testing laboratory, where they will use equipment not available to the general public. An earlier, popular test was called a Lauigram, which was an X-ray diffraction test taken of each individual pearl, in which concentric layers of crystals gave one result, while a pearl with a mother-of-pearl bead nucleus gave another result, due to the bead's parallel layers of crystals. The advent of non-nucleated cultured pearls, which have the same crystal formation as natural pearls, has resulted in this test being unreliable for the detection of natural pearls.

The most usual laboratory test on pearls today is X-ray radiography, which reveals an image of the inside of the pearl. It has the added advantage of being able to test whole strings of pearls at the same time. An indication is given below of some of the results that are likely to appear on the negative X-ray radiographs. They are possible because the organic matrix in pearls is radiolucent to X-ray, in contrast to the layers of calcium carbonate which are more radio-opaque.

- Natural pearls may show concentric rings all the way to the centre.
- In a natural pearl, any cavity in the centre should follow the contours of the pearl's shape.
- Non-nucleated cultured pearls may show the same rings as a natural pearl, but with evidence of an irregular cavity in the centre, caused by the mantle tissue graft.
- Nucleated cultured pearls should reveal a complete two-dimensional circle near the surface of the pearl, caused by the organic matrix between the mother-of-pearl bead and the nacre.
- The image of an imitation pearl will depend upon the nature of the bead used, as all the simulants have different radio-opacities.
- The negative image of a black pearl that displays white areas on the radiograph, either in arcs or as uneven patches, suggests a pearl that has been dyed black using silver nitrate. The white areas indicate where the silver nitrate has been deposited in high concentrations.
- The presence of organic dyes cannot be detected by X-ray radiography.

Conservation and availability

The vast majority of pearls on the market today are cultured, that is to say they are farmed, and therefore they should be in no danger of extinction. Problems can arise if there is an upset to the ecological stability at a pearl farm, as pearl oysters and mussels are susceptible to pollution, sudden changes of temperature, or the phenomenon known as red tide. This occurs when there is an explosive proliferation of certain tiny sea creatures, in numbers so great that the local waters cannot sustain them, so they die. The resulting toxins in the water are fatal to pearl oysters.

The famous Biwa cultured pearls, named after the lake where they were farmed in Japan, were almost completely exterminated in the 1970s and 1980s. This was caused by a combination of factors, which included pollution from agricultural waste, overfishing, and a freshwater form of red tide. It is taking years to restock the lake with the mussels needed to produce these pearls.

In the mid-1990s the Akoyas in Japan were struck by a virus that reduced cultured pearl production by 75 per cent, seriously damaging the cultured pearl industry of that country.

In Europe, the European freshwater mussel is facing extinction from overfishing. Once numerous, the mussels have declined in number to such an extent that they have completely disappeared from many of their former habitats. In several places, for example Scotland, there is now a complete ban on fishing for them.

The pearl mussels of the eastern United States have also suffered from overfishing, and some species are in danger of vanishing altogether. Most are now protected, but are still endangered by river management works and by pollution.

Past and present uses

Through the ages pearls have had two main uses: to be worn as jewellery, and to be sewn onto clothes. Occasionally they have been put to ornamental use as well.

The origins of pearls have baffled people for millennia. Some thought that they were eggs laid by the molluscs. The Persians believed them to be the tears of the gods. In China it was believed that they were caused by rain being caught in the shells, and that moonlight induced them to grow. The Greeks had a similar idea and thought that the dew from the moon entered the oysters as they swam on the surface by moonlight, and the dewdrops became pearls.

Pearls have been held in high esteem in most countries of the world, though in some, especially in South East Asia, they were originally not highly regarded. The mother-of-pearl shells were sought after, but any pearls found in them were discarded.

Pearls are not considered very durable, but, given the correct conditions, they can last forever. There is one example of a pearl found in a burial mound in the Middle East, which dates from about 2000 BC.

There are records of diving for pearls in 200 BC in the Arabian Gulf, and the Chinese have used freshwater pearls since antiquity. For centuries they traded for them, even obtaining them from Japan where pearls were not yet popular.

The Romans loved pearls and those of high standing used them as adornment on furniture, as well as on clothes or as jewellery. Nero had a throne covered in pearls, but Caligula gave his horse, Incitatus, a pearl necklace when he elevated him to consul. Caligula himself had pearls on his slippers. The Romans originally bought their pearls from the Persian Gulf. Later they used European freshwater pearls, and it has been suggested that it was the pearls in British streams that brought the Romans to this part of the world.

In Byzantium pearls were sewn onto clothes, and used to adorn church artefacts. Much the same was happening in Europe at this time.

Christopher Columbus was sent to the Americas with a list of items to bring back for the King and Queen of Spain. Topping the list was pearls. Columbus was not very successful in finding pearls, and only succeeded on his third trip, in 1498. His fourth trip – when he also returned without pearls – was his last. Within a couple of years another explorer sent out from Spain had found the islands that became known as the Pearl Coast, just off the north coast of what is today Venezuela. From then on, pearls – and slaves – were brought back to Europe by the shipload, and in 150 years the area was completely stripped of pearl oysters.

During this time the courts of Europe were awash with pearls from South America, and from their own rivers. Queen Elizabeth I of England certainly overindulged in pearls. She wore ropes of them that reached her knees, and it is said that she had 3000 dresses embroidered with them, but it is interesting to note that a lot of the pearls on her clothes were imitations.

India, in the mid-seventeenth century, was just waking up to the beauty of pearls and starting to use them extensively. Again, they were sewn onto clothes and used as jewellery, especially necklaces. Shah Jehan, who built the Taj Mahal, had a throne – the Peacock Throne – encrusted with gemstones, and with hundreds of pearls.

By the eighteenth century Europeans were using their pearls with a lot more subtlety, and in North America and Australia (both colonized by England), pearls were used in much the same way. In Russia they were still being sewn onto clothes and used to adorn icons.

A century later a totally different type of pearl had gained popularity in the west: seed pearls. They were tiny, natural, white pearls of less than 2 millimetres in diameter. Though they were worn in large quantities, the effect was decorative rather than overpowering. They were strung and woven on white horsehair to give a lacy effect, or threaded onto wire. Popular motifs were flowers and symmetrical patterns. Seed pearls were worn as chokers, bracelets and brooches,

Figure 9.23 Seed pearl necklace. English, nineteenth century. *Victoria and Albert Museum, London.*

and were also popular as hair ornaments, where they were often combined with fresh flowers. Motifs such as butterflies were often mounted 'en tremblant', that is to say threaded on wire in such a way that the butterfly would tremble as the wearer moved (Fig. 9.23).

The diving methods used to retrieve pearls in the Arabian Gulf remained the same for centuries. Young men, at one time mostly slaves, were taken out on boats from which they dived. They could hold their breath for some minutes, and descended to a depth of 20 or 30 metres – possibly with the help of a nose clip – where they collected a few oysters and returned to the boat. The exercise was repeated again and again with little thought to the divers' safety, and the mortality rate was very high. Diving for pearls in the area continued until the advent of cultured pearls in the twentieth century. Possibly more importantly for that region, oil had also been discovered in the Gulf.

Natural pearls from the South Seas were first imported to Europe in 1845, and in 1881 the huge silver-lipped oysters were discovered in the seas around north-west Australia. In about 1900 the American freshwater pearl mussel industry started up. It was concentrated around the Mississippi, and began with harvesting mother-of-pearl, most of which went to the button industry.

Pearl cultivation is a much older science than is generally realised. The Swedish botanist Carl Linné had attempted, with some success, to culture blister pearls in the eighteenth century. And long before that – some say as early as AD 500 – the Chinese had been culturing blister pearls, using half spheres or images of the Buddha as the nucleus.

Culturing of round pearls began in Japan. A man by the name of Kokichi Mikimoto developed a method of completely wrapping a bead nucleus in mantle tissue and implanting it in an oyster, where it developed into a pearl. Very soon after that, another man, called Tatsuhei Mise, discovered that it was only necessary to insert a sliver of mantle together with the bead nucleus, and that the oyster would wrap the mantle around the bead itself. This was obviously much quicker and easier, and is basically the method still used today, though with some refinements. The two gentlemen decided to join forces, and in 1916 Mikimoto patented the method.

Although at first very unpopular and shunned as artificial, cultured pearls soon became popular and took over from natural pearls. Without careful examination and testing, it was often not possible to see the difference, so people were no longer prepared to pay for natural pearls when the cultured ones looked as good.

In Australia there was a valuable trade in mother-of-pearl for the button industry, using the large *Pinctada* oyster shells found off the north-western coast. With the advent of plastic buttons in the mid-twentieth century, the shell trade ceased, and another use had to be found for the shells. It was not long before Australian cultured South Sea pearls, cultured in the *Pinctada* oysters, appeared on the market.

Natural black cultured pearls were successfully produced in French Polynesia in the 1970s. These types of pearls are expensive to produce as, for example in Australia, the process involves scuba divers with boats equipped as laboratories, to tend to the oysters that are kept off-shore. By comparison, the Chinese freshwater mussels lie in shallow streams and lakes where it is possible to wade out to them, and the floats holding their cages are often made of empty plastic drinks bottles. The value of a single Chinese pearl today is a far cry from the value of one particularly special pearl in Roman times, which, it is said, paid for a whole battle.

Through the ages there have been a few pearls that have become famous. One such pearl is La Peregrina. It is a white, pear-shaped pearl of exceptional size and beauty. It may have been fished from near Panama, or it may have come from the Pearl Coast, and legend tells that the young slave who found it was granted his freedom as a reward. The pearl belonged to Philip II of Spain, and was later given by Ferdinand V to his fiancée, Mary Tudor. It passed through many royal and noble hands, including those of Josephine Bonaparte, until it was eventually sold to the Marquis of Abercorn in England. The last record of the pearl states that it was bought at auction by the actor Richard Burton as a Valentine's Day gift for Elizabeth Taylor.

10 Shell

A shell is the calcareous outer part of a mollusc's body. Molluscs are animals living both on land and in the sea, but the shells that are used as gem material come mostly from marine or freshwater molluscs, though some land-based snails are also used.

Any shell that is sufficiently solid and attractive can be used for decorative purposes. Those with nacreous interiors, or with a structure made up of layers of different colours, are the most commonly used.

Structure and properties

Molluscs vary greatly in size, ranging from minute creatures of less than 1 millimetre across, to giant clams. Most, but not all, have shells.

Their bodies usually consist of a muscular part, sometimes a head and a foot, plus a soft, non-muscular mass that remains inside the shell. Attached to this is an envelope of tissue that lines most of the interior of the shell, called the mantle.

A mollusc's shell is formed by the mantle, which deposits an organic matrix (sometimes called conchiolin), and calcium carbonate, which forms crystals. The mollusc absorbs the chemicals necessary for this from its surroundings.

Most shells are laid down in three or four layers. There is an outer covering of organic material, called the periostracum. This is waterproof, and may be smooth or horny. Inside the periostracum are two or three layers of calcium carbonate, called the outer, middle (where appropriate) and inner shell layers.

The calcium carbonate crystallises in different ways, according to the species. The crystals may be prismatic, columnar or in platelet form, and each layer may be different. In some shells the layers also differ in colour. In all these layers the crystals are held together by a fine web of organic matrix.

Some shells have a porcellaneous inner layer, while others produce a nacreous layer. The nacreous effect is called 'mother-of-pearl' and is caused by tiny, overlapping platelets of calcium carbonate in the form of aragonite, which disperse the light to give a play-of-colour. The platelets are polygonal and lie horizontal to the surface. In gastropods and nautilus they are stacked like coins, but they can also be alternated, like a brick wall, as is the case in most bivalves. There can be more than one nacreous layer, as in, for example, freshwater mussels.

In non-nacreous shells the inner layers are composed of thin, sheet-like plates or lamellae, which can either lie perpendicular to the shell's surface, or in blocks which can alternate in opposite directions at an angle of 45 degrees to the surface.

The periostracum, outer and middle layers of shells are secreted by the edge of the mantle, so that they increase more in size than in thickness as the animal gets older. The inner layer is secreted by the whole surface of the mantle, and therefore gets thicker with age. However, if the shell is damaged, this surface is capable of producing the appropriate materials to repair all three layers.

The species

Molluscs form the second largest phylum in the animal kingdom, with around 128 000 extant species. They can be divided into six groups, but only three contain species providing gem or ornamental material: the gastropods, bivalves and cephalopods.

Gastropods

Gastropods are univalves, that is to say they have one shell. The shell spirals to give it strength, and the spiral is usually equiangular, which means that it keeps its shape as it grows larger, allowing for undisturbed growth. Gastropods have eyes and they crawl by the use of one, flattened foot. This group includes conch, helmet, abalone and paua, top and cowrie.

The **queen conch**, *Strombus gigas*, is used mostly for cameos or beads. It has a brownish-yellow periostracum which is thick and horny and which is usually removed completely. The middle layer is white and the inner layer is typically pink, though in younger animals it can be a strong orange-pink (Fig. 10.1). The aragonite crystals form a striated appearance, which may appear as a flame pattern. Each layer's striations lie in the same direction, which is a characteristic that helps to distinguish this shell from a helmet shell. The pink colour fades with exposure to light, finally becoming pure white.

Figure 10.1 Conch shell, one end removed to show cross-section.

The **helmet shell** has many varieties, two of which are commonly used as gem material and carved into cameos. As with the conch, the aragonite crystals give a striated appearance, but each layer is at right angles to the next. The inner, coloured layer displays the same flame pattern as do conch shells. The outer layer is sometimes incorporated in the carving, sometimes completely removed. Only one area of the shell provides good-quality cameos, while the rest is used for less expensive souvenir-style items.

The **black helmet**, *Cassis madagascarenis*, comes from the Caribbean, (though it was originally thought to come from Madagascar, hence its Latin name). It has a chalky-white outer layer, a white middle layer, and a chocolate brown inner layer.

The **bull mouth helmet**, *Cassis rufa*, has a brown outer layer, a creamy middle layer, and a tawny-red inner layer.

One other helmet shell, the king helmet, *Cassis tuberosa*, is occasionally used instead of the black helmet. It has the same colouring.

The **cowrie**, of the genus *Cypraea*, has over 300 different varieties. Some are tiny and are used as beads, some are pure white all over, and some have a typical, dappled appearance. They come mostly from the

Figure 10.2 Cowrie shells: natural and worked.

Indo-Pacific region. They consist of a whorl that envelopes the whole body. The whorl is only visible in young cowries as in an adult it is covered by the lip of the shell. The most popular variety of cowrie is the **tiger cowrie**, which is sold in its natural state, or can be made into all manner of souvenirs, including those with crude cameo carvings on the shell. The periostracum is invisible on mature cowries, and the outer layer is a thin but shiny mottled brown, which does not need polishing. The middle and inner layers have a bluish hue (Fig. 10.2). Another popular version is the **money cowrie**, which is much smaller and is used as jewellery.

Abalone and paua are shells of the Haliotis family which have a nacreous inner layer. Haliotis shells occur worldwide, though the larger species grow in temperate waters. The temperature of the water affects the colour of the shell. Those with a predominantly greenish hue have much thicker shells, with up to ten times the number of aragonite platelets compared with the bluish shells from cooler waters. The shell is open, thick, and made of a single whorl that grows outwards at the same level, giving the shell its relatively flat form. It has a row of holes along one side. The animal moves little and tends to cling to one spot.

The molluscs from New Zealand are known as paua, while their cousins across the Pacific on the American coastline are known as abalone. Both occur in glorious iridescent greens, blues, pinks and purples. The nacreous layer has a surface pattern resembling waves, which is caused by concentrations of organic matrix. In the centre is a more uneven area where the animal was attached to the shell. The

Figure 10.3 Abalone shells: natural, polished and cabachons.

Figure 10.4 Trochus shells: natural, polished, and mother-of-pearl buttons.

play-of-colour is caused by light diffraction through the aragonite platelets that make up the inner and middle shell layers. The outer layer – which is rough and a dull beige or brown – can be removed to reveal the nacreous layer immediately beneath. Shells are often sold whole with the outer layer removed, or cut up and made into souvenirs or jewellery (Fig.10.3).

Top or **trochus** shells, from the family Trochidae, are found in the Indo-Pacific region near coral reefs. The **cone-shaped top, mottled top** and **commercial trochus** have nacreous inner layers and are used in the mother-of-pearl button industry. The shells are triangular, cone shaped, with a flat base and flattened whorls. The outer shell layer of the commercial trochus is brown and white, while the others vary in colour from red to pink and white, sometimes with splashes of brown, grey, or green (Fig. 10.4).

Opercula. Most aquatic gastropods have a round piece of shell – called the operculum – on the back of the foot, which is used as a form of trapdoor when the mollusc retreats into its shell. It is secreted by a group of cells on the back of the foot, where it rests while the animal is moving. It may be made up entirely of protein, but can also be made of aragonite or calcite. Opercula from turban shells are used as unusual gemstones, and are made of aragonite. They have a whirled appearance on the back, and the front often displays circles of blue, green or brown colour on a white background. In their natural state the front is covered in tiny round bumps, but these are polished away (Fig. 10.5).

Bivalves

Bivalves have no head and usually no eyes, but have two, symmetrical mantle flaps which enclose the body and which each secrete a shell. The two shells are hinged together and can be closed by muscular movement. Some bivalves remain in one spot, cemented to a firm base, while others are mobile. The group includes oysters and clams.

Pearl oysters are dealt with in greater detail in Chapter 9 ('Pearl'), as the shells have a great influence on the pearls they produce. Nowadays they are regarded as less important than the pearls, though originally the pearls were regarded as a by-product of shell harvesting.

The shells are beautiful in their own right and can be used in a variety of ways. The best known are the large silver- and golden-lipped oysters, the black-lipped oyster and the akoya (Figs 10.6 and 10.7).

Some of the freshwater mussels that produce pearls also display attractive shells. The cultured pearl industry is very dependent upon beads made from freshwater mussel shells (Fig. 9.10), notably the washboard and pigtoe mussels, which are used for nuclei for cultured pearls.

Clams vary in appearance and size and belong to different families. They seldom have a nacreous inner layer and do not have particularly

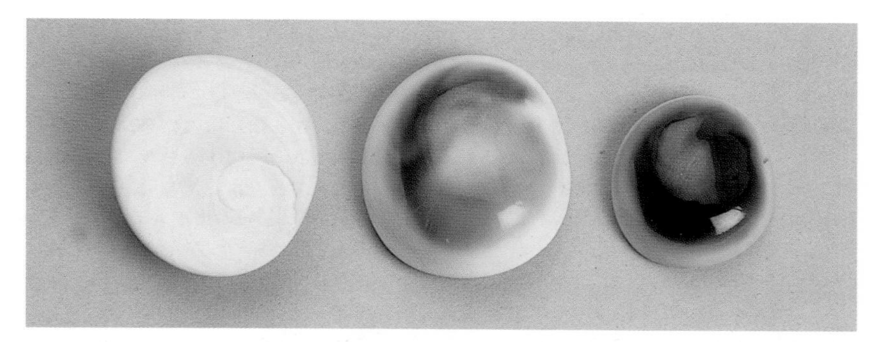

Figure 10.5 Three opercula (magnified).

Figure 10.6 Mother-of-pearl brooch.

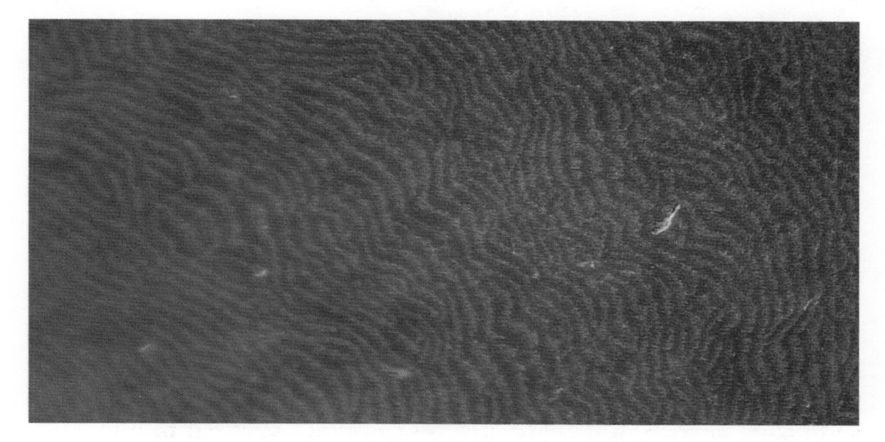

Figure 10.7 Mother-of-pearl under high magnification, showing overlapping aragonite platelets.

attractive colours. On rare occasions a clam produces a pearl, though this is not nacreous. The group is mentioned because the shell of the huge **giant clam**, *Tridacna gigas*, has occasionally been used as a decorative or functional item, for example as a baptismal font.

Scallop shells, of the family Pectinidae, though once very popular, are now rather out of fashion. They come in a variety of colours but the inner layer is usually a dull cream and is not nacreous.

Figure 10.8 Coque-de-perle (magnified).

Figure 10.9 Nautilus shells: natural and polished, cross-section, slices and coque-de-perle.

Cephalopods

Cephalopods are mostly intelligent creatures and can move quickly through the water. They are carnivores. Their shell is of less importance than those of other molluscs and some cephalopods, for example squid, have evolved to carry their shells inside their bodies.

The **nautilus**, *Nautilus pompilius*, has changed little since its ancestors were around several million years ago. It is found at a depth of about 250 metres in warm seas of the Indo-Pacific region. It has a shell that is spirally coiled, with a creamy coloured periostracum with brown, zebra stripes. The centre of the whorl is a nacreous silver-blue which, when used in jewellery, is known as the 'coque-de-perle' (Fig. 10.8).

The shell is partitioned into connecting chambers, with the mollusc occupying only the last one. The rest are filled with gas to give buoyancy. The inner layer of the shell is nacreous, and, as with paua shells, the periostracum can be polished away to reveal the nacreous layer beneath (Fig. 10.9). The material is most commonly used as mother-of-pearl inlay.

Other shells

There are many more shells than those named in this chapter, which are used in jewellery, or for decorative and ornamental purposes, in various parts of the world. As already stated, any mollusc that is robust and pretty enough can be used. One such example is the **spiny oyster**, which is a native of the seas around Panama. It is a bivalve belonging to the family Spondylidae, and has been used in native American jewellery. The shell is covered in long, slightly curved spines. The inner layer is white with a red lip, while the outside is red or reddish-brown, with the spines fading to white at the tips. The spines are removed when the shell is worked, leaving an attractive, highly polished, red-streaked surface (Fig. 10.10). In New Zealand another bivalve, commonly known as the **kuku**, is polished to make attractive jewellery. It is also pink. The so-called **window-pane oyster**, *Placuna placenta*, has a thin shell, which is translucent white, and is used to make hanging mobiles or small boxes as well as inexpensive jewellery. In times past it was used for window-panes in the Far East – hence its name.

TREATMENTS AND USES

- All shell can be cut or etched. The nautilus is an example of a thin shell that was once popular for use as ornamental cups, with sections cut away to form a pattern and a picture etched and stained onto the remaining surfaces.

Figure 10.10 Spiny oyster shells: natural and polished.

- Intricate carving and cutting of nacreous shell into fretwork has been used for many articles such as boxes and fans.
- Shells of sufficient thickness, which have layers of different colours, can be carved into cameos. The periostracum, and possibly the outer shell layer are removed, so that the top layer of the carving utilises the white middle layer of shell, and the coloured inner layer of the shell provides the background for the cameo. Sometimes the outer shell layer is left as an integral part of the design (Figs 10.11 and 10.12).
- Very thick shell can be cut and made into beads. Beads from shells with alternating colour layers (such as the conch), or with nacreous

Figure 10.11 Conch shell cameo (magnified).

Figure 10.12 Helmet shell cameo (magnified).

layers, are used to make attractive, inexpensive beads (Figs 10.13 and 10.14). Shell beads can also be dyed to simulate, for example, black pearls.

- The shells of certain freshwater mussels are used as the nucleus in marine cultured pearls (Fig. 9.10).
- Shells can be dyed, though this is usual only with abalone or paua shells where a dye enhances or emphasises the colour already there, or with tiny, thin shells sold as inexpensive jewellery (Figs 10.15 and 10.16).
- To protect the surface of a shell item, it can be capped with plastic. The resulting doublet is extra shiny and the subtle, natural sheen

Figure 10.13 Shell beads, showing layers.

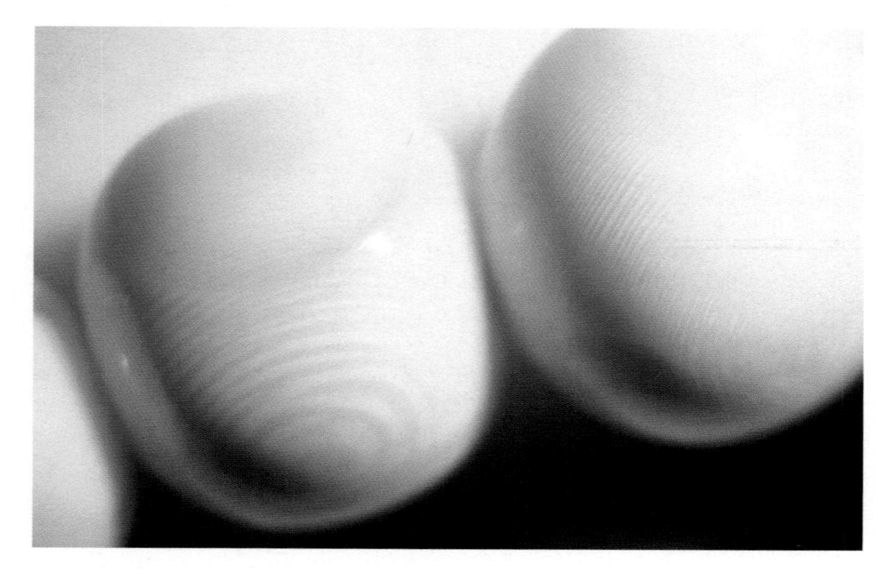

Figure 10.14 Detail of shell beads, showing structure.

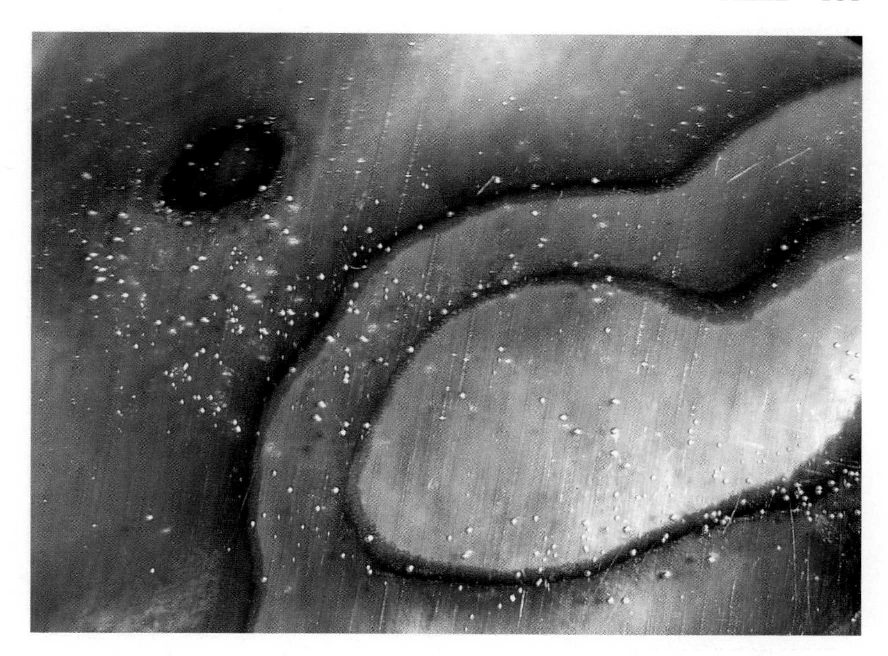

Figure 10.15 Detail of abalone shell showing accumulation of dye, and air bubbles.

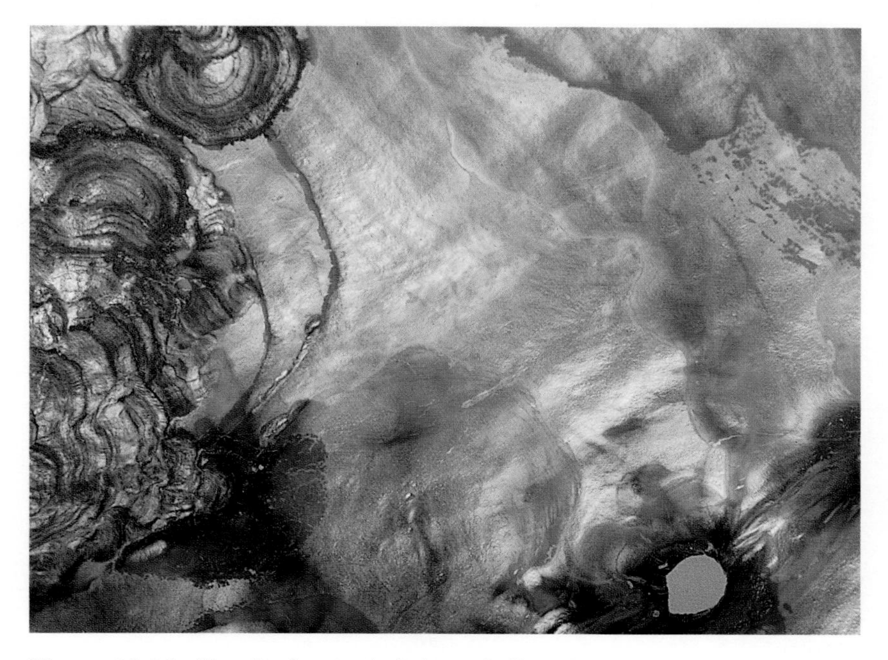

Figure 10.16 Detail of natural abalone shell.

of the shell is lost. It is usually only seen on small, inexpensive items such as dyed abalone or paua cabochons (Fig. 10.15).

- Similarly, thin pieces of shell such as paua and abalone may also be backed with plastic, forming a triplet.
- Small shells, or pieces cut from larger shells, can be incorporated in jewellery and used to adorn dresses, bags, saddles and bridles or even furniture (Fig. 12.2).
- As an inlay, pieces of thin nacreous shell are used, for example from nautilus shells. They may be etched and stained to enhance the pattern, or to give the impression that the inlay is made up of more pieces and therefore more intricate (Fig. 10.17).
- The grey-blue nacreous centre of the whorl of nautilus shell is used as a natural cabochon. It is called the 'coque-de-perle'. It is cut out and backed to give it body (Fig. 10.8).
- Mabé pearls are cultured blister pearls, produced against the shell of a mollusc. They are dealt with in detail in Chapter 9 ('Pearl') as they are used and sold as pearls (Figs 9.2 and 9.3).

Simulants

Shell is rarely copied as it is plentiful and cheap, though the more expensive items such as cameos which involve a lot of hand crafting

Figure 10.17 Mother-of-pearl inlay on box lid, etched and stained for greater effect.

have, inevitably, been faked. Mother-of-pearl has also been copied, probably because natural mother-of-pearl cannot be moulded, which limits its uses.

- **Coral.** Though not actually a simulant of shell, coral bears a resemblance, especially in carved or bead form. The structure of the

Figure 10.18 Two-tone moulded plastic, imitating shell cameo.

Figure 10.19 Detail of a plastic cameo, showing signs of moulding.

material seems similar, as does the feel. Careful examination will show a different structure. Under high magnification shell displays very much finer striations, which are lacking in coral.

- **Plastics** have been used to copy expensive shell items. In order to imitate shell cameos, a plastic substance can be moulded in layers of different colours. The results tend to be obvious and crude, with mould marks and air bubbles, lack of structure or marks from the carving tools, and a flat reverse side (Figs 10.18 and 10.19).

  Cameos can also be moulded in one colour of plastic and painted, to imitate the different coloured layers of the shell. This is usually done in creamy-white plastic with just the background painted an appropriate colour. These imitations are easy to spot (Fig. 10.20).

  A third method of making plastic cameos is by gluing layers of different coloured plastic together. It is highly unlikely that two or more pieces of shell could be joined in this way as each piece will have its own natural curve that may not necessarily match the next piece.

  Plastic imitations of mother-of-pearl are more tricky to detect. One simulant is made by suspending mica in a clear synthetic polymer. The effect is very realistic with an apparent play-of-colour, but the shape is often a give-away as the material is moulded. For example, a round napkin ring with no joins in the material could not be made from one piece of shell. Small items such as buttons are more difficult to identify.

Figure 10.20 Painted plastic, imitating shell cameo (magnified).

- **Glass and porcelain** have occasionally been used to imitate shell, but they are unconvincing. Both glass and glazed porcelain have a vitreous lustre that is different in appearance to that of the natural material. They also feel cold and hard, and have a complete lack of structure. Both materials are likely to show signs of having been moulded.

TESTS AND IDENTIFICATION

Shell is made of calcium carbonate – a mineral – and held together by an organic matrix. As with all other organic gem materials it is rarely possible or advisable to carry out any of the usual gemmological tests which involve chemicals, as they could damage the material. Visual identification is the safest option and should be sufficient.

Visual identification

- Viewed with a 10× lens, non-nacreous shell displays a structure of faint striations. The orientation of the striations in each layer of colour is approximately perpendicular to the next layer. The exception to this rule is conch shell, which has striations in the same orientation (Figs 10.21 and 10.22).
- Conch shell fades with exposure to light. This can result in an item that is pure white and has striations running in the same direction. In small items such as beads, this can resemble white coral. When viewed from various angles the structure of coral becomes apparent.
- Viewed from the back (the inside surface of the shell), cameos may display the flame pattern caused by the structure of the shell, as the striations bend slightly at the surface.
- Nacreous shell should be viewed under a microscope at high magnification to see the minute, overlapping platelets of aragonite from which it is formed (Fig. 10.7).
- Mother-of-pearl buttons can often be identified with the naked eye by looking at the back of the button, which usually retains some of the coloured layers of the trochus or other shell from which they are made.
- Pure white mother-of-pearl buttons, made from freshwater mussels, may have to be viewed under a microscope to see the aragonite platelets. Lack of these indicates a plastic imitation.
- The shape of an item gives an indication of its origins. A cameo with a slightly concave back suggests shell, whereas a totally flat back may be an indication of another material.

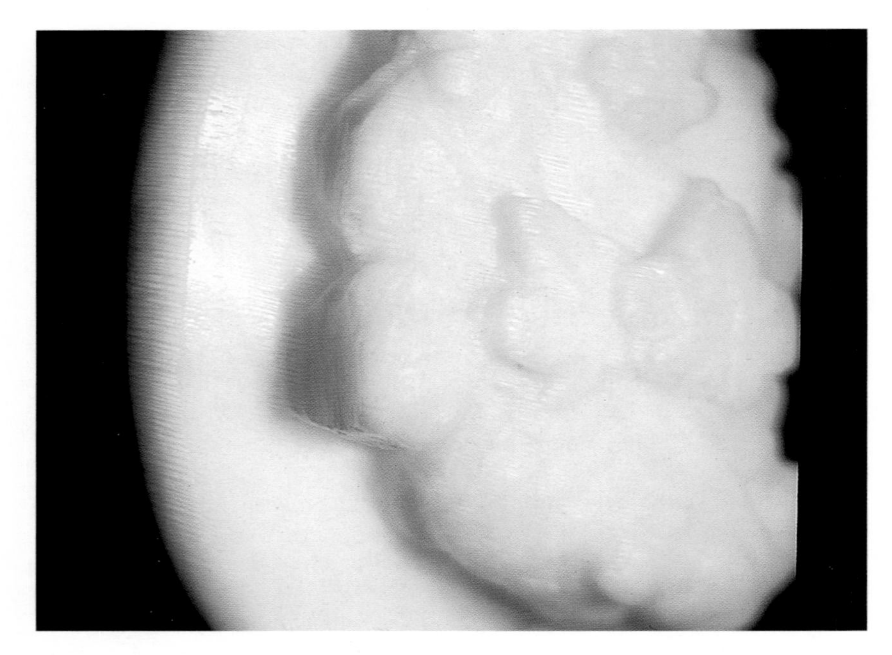

Figure 10.21 Detail of slightly faded conch shell cameo, showing striations running in the same direction.

Figure 10.22 Detail of helmet shell cameo, showing striations at right angles to each other.

- Similarly, a round napkin ring made from one piece of 'mother-of-pearl' suggests a moulded simulant, as no shell would be thick enough for this shape.
- Dyed shell may have an artificially bright colour. Viewed with a 10× lens the dye can be seen to accumulate in patches. This is especially noticeable in the areas of organic matrix on the surface of dyed abalone or paua shell (Fig. 10.15).
- A clear plastic cap forming the top section of a doublet tends to trap air bubbles, which are clearly visible with a 10× lens (Fig. 10.15).
- Shell is cold to the touch. Porcelain, glass and coral also feel cold to the touch but plastic is warm.
- Shell is heavier than plastic, but feels similar in weight to porcelain, glass or coral.
- Total lack of structure in the material indicates a man-made simulant.
- Air bubbles, mould marks and total lack of sharp corners from cutting indicate a moulded simulant (Figs 10.18–10.20).
- Cameos made in separate sections or layers indicate a simulant. Doublets of ivory pictures on a coral background have been produced, but they are rare, and are recognisable under careful examination.
- Swirls of colour are likely to be an indication of a moulded simulant. However, it should not be forgotten that swirls of colour can arise in the different coloured layers of some shells, when seen in cross-section. They are unlikely to be present in cameos as the coloured layers are carved, but they may appear in beads.
- Glass and porcelain have a high, vitreous lustre while shell generally has a softer lustre.

Tests

- **Ultraviolet light.** Some shells – for example, mother-of-pearl – fluoresce a chalky-white under UV light. In conch cameos the white area has a stronger fluorescence than the pink, and in helmet cameos the brown or orange layer can be inert. With the exception of casein, plastics are inert.
- **Acid test.** A drop of 10 per cent hydrochloric acid will effervesce if it comes into contact with shell. However, it will have the same effect on coral. This is a very destructive test as it will leave a hole in the surface tested.

Conservation status and availability

Attractive shells are plentiful and can be collected on any seashore, but the more specialised types that are used for gem or decorative purposes are not so readily available.

As with so many animals the conservation status of the various species changes from year to year. Nautilus has been thought to be endangered but is still widely sold, while some of the freshwater mussels, from which mother-of-pearl nuclei are made for the cultured pearl industry, are in real danger of extinction and are now protected.

Many molluscs are very slow growing and take a long time to reach maturity, so overharvesting is extremely detrimental to the stocks as they take so long to recover. Molluscs are also very susceptible to pollution, and the population of whole areas can be killed off quickly by a change in the chemistry of the water in which they live. This can be caused by events such as climate change altering the algae content of the sea, or by chemicals such as fertiliser seeping into the rivers (see also Chapter 9, 'Pearl').

Past and present uses

The use of shells goes back to prehistoric times when molluscs were caught for eating and their shells used as primitive implements, or for adornment. They have been popular through the ages in all parts of the world. The larger shells of the warmer climates have generally been the most sought after, yet still people delight in finding a tiny, perfect shell on any beach. Even the shapes of shells have frequently been copied. Shell motifs can be seen as decoration on buildings, and brass wind instruments have copied the expanding spiral cavities of gastropods' shells.

Cowrie shells from the Red Sea have been found in Stone Age tombs in northern Europe, which would indicate that they were used for trading. In Sumeria in about 2500 BC, mother-of-pearl was combined with lapis lazuli and mounted in silver to make head-dresses. The Aztecs and the Incas, although they had access to precious stones, valued mother-of pearl and used it, combined with turquoise and jade, as inlay.

Shells were used as coinage in many countries. In North Africa and in Asia a certain type of small cowrie was used, hence their name: 'money cowries'. Native Americans used quahog shells, which they cut and threaded as bead currency called 'wampum', and until relatively recently, mother-of-pearl was used by the people of New Guinea in place of money.

In Africa almost any type of shell – including those of land snails – has been used as jewellery, and strung like beads to make necklaces, bracelets or anklets. Shells are still a popular form of adornment, and are often combined with other organic materials such as feathers, or with brightly coloured glass beads.

In some places shells have not been much admired by the locals, and have been regarded as worthless. One such place is Tahiti, but with the arrival of the Europeans in the early nineteenth century, the local people discovered that they could sell their shells. An industry developed which subsequently employed thousands of Tahitians who depended upon it for their livelihood.

Some shells have been used as ceremonial objects. The heavy chank shell from the Indian Ocean found use as a trumpet, and was blown at weddings or when going into battle. The people of the Mediterranean and the Pacific used their local shells, the trumpet tritons.

Giant clams can reach almost a metre in diameter and are very solid with a porcelain-like inner layer. Mounted on plinths they have been used in churches as benetiers (holy water fonts), and examples exist in stately homes and palaces where they have been used purely for decoration.

As with all the other organic materials, shells were involved in myths and legends. The renaissance painter Boticelli depicted the goddess Venus being blown into shore standing on a huge shell. Some shells were said to guard against the evil eye. Opercula have been placed in temples as eyes, to keep a watch over the building and the worshippers.

The relief carvings – called 'cameos' – have been worn for at least the last 2300 years. For centuries they were carved in hardstone, especially banded agate where the different coloured layers were used to advantage in the carving. When shell was discovered, it was an obvious substitute for hardstone as it was much easier to carve, and soon became more easily obtainable. The first shells carved were possibly cowries, but these were quickly followed by helmet and conch shells, which were thicker. As with banded agate, there were coloured layers that could be incorporated into the pattern. The first shell cameos were probably produced around the fifteenth to sixteenth centuries. Portraits have remained a popular motif, but in later times, for example in the late nineteenth century when cameos were especially popular in Europe, romantic images of pretty ladies and cherubs with flowers became equally popular.

Early on, conch shells brought back to Europe had been ground down and used as a component of porcelain. As the shell carving industry evolved, various centres for the craft developed. The most famous centre is Torre del Greco near Naples in Italy, where shells are still made into cameos, and coral is still carved.

At the beginning of the seventeenth century the Dutch brought nautilus shells back from the Dutch East Indies and perfected the art of cutting and carving these in delicate and ornate designs, sometimes with pictures added that were etched into the shell and dyed with

Figure 10.23 Bridal fan. Lace and mother-of-pearl.

indian ink. Mounted on silver or silver gilt stands, they were extremely fragile and used purely for decoration.

In Europe shell jewellery consisted mostly of conch and helmet cameos, made into brooches, pendants, ear-rings and occasionally bracelets (though these would have been fragile). Mother-of-pearl from nacreous shells was also used, and could be intricately carved and fashioned into the same sort of items. Nacreous shells proved even more versatile than porcellaneous shells, because many of them were almost flat and therefore could be made into objects such as hair ornaments, fans, small boxes, gaming pieces and all sorts of trinkets (Fig. 10.23).

For several centuries mother-of-pearl has been used as an inlay material. This was particularly popular in the Far East, where items of lacquer or tortoiseshell, as well as wood, were inlayed with mother-of-pearl, often using nautilus shell. In India mother-of-pearl has been used as inlay on large pieces of wooden furniture and caskets, and as an overlay on brass items such as jugs and basins.

At one time the mother-of-pearl button industry was huge; indeed it was so big that it almost wiped out entire populations of freshwater mussels in North America. The industry was in turn almost killed off by the advent of plastics. Today some buttons are still produced in the Far East and are made mostly from trochus shell. A few more exotic – and expensive – examples from other shells are also manufactured.

The shell industry is still very active in various parts of the globe, especially in the Far East, where every souvenir shop is stocked with items made of, or using, shell. In New Zealand paua shell is made into anything from beads to boxes, and mother-of-pearl and paua is made into laminates for wall coverings. Cameos are not only made in Italy, but also in Japan. Most of them are now mass produced, albeit by hand, for the tourist trade. In northern Europe it has long been a custom in every seaside resort to sell trinkets covered with shells, and shell necklaces are sold to tourists worldwide.

11 Coral

Today, coral is the name given to sea dwelling, soft bodied, carnivorous animals with polyps, most of which live in colonies. They belong to the phylum Coelenterata. There are thousands of different corals throughout the world's warmer seas. Coral is also the term used to describe the skeleton of the polyps, which may, in some cases, be used as a gem or ornamental material.

To be precise, the word 'coral' has no formal meaning in taxonomic science, but was originally used to describe any organic material which grew attached to the sea floor, and possessed a hard skeleton. Thus it covered calcareous algae and other creatures, as well as coelenterates – some of which are the animals today considered to be 'corals'.

Structure and properties

Most corals prefer warm waters, though those growing at great depths are in cooler water. Corals have a multitude of growth patterns and shapes, some of which can vary within a species which adapts to the local conditions such as currents. They attach to hard surfaces and some types form reefs – the Great Barrier Reef off the north-east coast of Australia is the most famous of all. Most corals used for gem purposes are not reef forming. An exception is *Heliopora*, or blue coral.

Coral polyps have a similar appearance to sea anemones, to which they are related. They are usually cylindrical, and vary in size from 1 millimetre to several centimetres in diameter, according to their species. They are biologically simple animals, basically consisting of a mouth that is surrounded by a ring of tentacles at the top of the cylinder, and a gastrovascular cavity. They have no head. In some corals each polyp is connected to its neighbours by more living tissue, which is called the coenosarc. In other corals the polyps are so close together that the coenosarc is absent. The undersurfaces of the polyps and, where present, the coenosarc produce minute particles of hard material. In the majority of corals the particles fuse to form the coral's skeleton underneath the living tissue, though in one group – the 'soft corals' – this does not happen. For the majority of corals in this group, the particles remain separate with the result that the coral lacks rigidity.

The minute particles may be made of calcium carbonate, or of a horny material. In some corals this material is gorgonin, in others antipathin or chitin. All three are closely related to keratin (which is widespread in the animal world, as a component of skin, fur, fingernails, and so forth). A few corals have a combination of both calcium carbonate and horny material. Corals with a skeleton made of horny material are slightly less rigid than those with a skeleton made of calcium carbonate, though this is not noticeable when the material has been worked.

In some coral types, a mass of basal nutrient canals connects the polyps. The corresponding grooves in the skeleton run longitudinally along the branches and give the skeletal material its typical, ridged surface pattern. In these types of coral it is normal also to see a central canal running longitudinally through the centre of the skeleton. This appears as a central spot in a cross-section of the material.

The skeletons of other corals have a more spongy appearance, punctuated by a mass of holes. In these, the polyps have occupied the surface holes of the aragonite skeleton, and move further out as the colony grows and new layers of aragonite are laid down.

Many coral skeletons, especially those of reef corals, are white. Other corals have coloured skeletons, some of which have been used in gem and ornamental material. A large number of the living corals viewed in the sea are coloured by their living tissue, and not by their skeletons. Some of this colour comes from minute algae that inhabit the living tissue of the corals. Their relationship is symbiotic: the algae use sunlight to produce sugars by photo synthesis, which are needed by the corals, and the algae live off the corals' waste products.

The species

Many years ago the term 'coral' was generally regarded by the jewellery trade as applying to only one species, which was called precious coral. It was red in colour, and took a very high polish. Today there are many gem-type corals on the market, and they come in a variety of colours and textures. Occasionally types of coral that have previously been unused also appear on the market.

It can be very difficult to know exactly to which species worked corals belong, as they are almost impossible to identify precisely in cut and polished form, and because few dealers know the names of the corals they are selling. Classifications can change, and different people use varying terminology. For example, a coral called a stony coral by some is called a soft coral by others. Further, the term 'precious coral' is used in the gem trade to signify corals of the genus *Corallium*, especially the red *C. rubrum*, whereas any coral that can be used for gem purposes is termed 'precious coral' in the scientific world. The following endeavours to distinguish the corals, but it should be noted that it is intended as a guideline to the most common gem corals, and not as a complete list.

Most of the gem corals belong to the subclass Octocorallia (previously called Alcyonaria). The polyps of all octocorals have eight tentacles. Another subclass, Ceriantipatheria, includes the horny corals, and the subclass Zoantharia includes fossilised corals.

In the rest of the chapter, the word coral will refer to skeletal material of coelenterates rather than the living animal, unless otherwise stated.

Corallium corals

The genus *Corallium* includes the porcellaneous red, pink, peach and white corals (Fig. 11.1). They all display very fine striations that run longitudinally along the branches, spaced between 0.25 and 0.5 millimetres apart. This spacing is characteristic for the genus. *Corallium* corals take a very high polish. The skeletons are of calcium carbonate, and have a branched form, looking like small trees (Figs 11.2 and 11.3).

The best known is *C. rubrum*, which was earlier called *C. nobilis*. This is the original gem coral, and has been prized for centuries. It is an even, soft, pink to crimson colour. It grows in the Mediterranean, where the best quality is said to come from around Sardinia. Closely related corals from other regions are now also used, as follows:

- *C. secundum* is a Pacific coral. It occurs in white to pale pink.
- *C. konojoi* comes from around Japan and the Philippines. It is white with occasional pink spots.
- *C. elatius* also comes from around Japan and the Philippines. It is reddish-orange, or pink.
- *C. japonicum* comes, as the name suggests, from around Japan. It is dark to very dark red.

Blue coral

Blue coral, *Heliopora coerulea*, has a very different structure, though it is also an octocoral. It has a massive, aragonite skeleton, which takes the form of fingers, or of plates or baffles, according to the water currents in which it grows, forming a large, compact bush shape (Fig. 11.4). It has tiny holes all over the surface, and appears porous, dry

Figure 11.1 Necklaces of *Corallium* corals.

and chalky. Its natural colour is a grey-blue, but the material is frequently colour enhanced with dye, and usually impregnated with a resin when it is worked, which gives it a slightly shiny finish (Figs 11.5 and 11.6). Although the skeletal material is blue, the living tissue is pale brown, so that the coral looks brown when it is growing.

Red soft coral

The name 'soft coral' is very misleading, but the red coral, *Melithaea ocracea*, that is now so popular for jewellery is of the order Alcyonacea.

Figure 11.2 *Corallium rubrum* rough.

This order covers soft corals, but also includes some that have a rigid skeleton. In *Melithaea* the minute hard particles produced by the living tissue have fused to make a hard aragonite skeleton. The skeleton looks somewhat spongy and brittle, rather than solid, and its surface is abrasive rather than smooth. It is usually an orange-red colour – though this can vary slightly – with intervals of mid-brown, which give it its typical, variegated appearance. It grows in a three-dimensional tree-like shape, in Indo-Pacific regions (Fig. 11.8).

Melithaea is often dyed a darker red, and, when used for jewellery, is always impregnated with some form of resin to give a smooth and

Figure 11.3 *C. rubrum* rough, detail, magnified, showing striations.

slightly shiny surface. It is not necessary to impregnate it for carvings (Figs 11.9 and 11.10).

Bamboo corals

These corals are so named because the jointed growth form of the branches bears a resemblance to canes of bamboo. They belong to the genus *Isis*. They are a combination of calcium carbonate and gorgonin, the longer calcareous sections being interspersed with nodes of dark brown gorgonin. The calcareous areas have similar, though coarser, striations to *Corallium* corals, with the lines about 1 millimetre apart. They occur in white or pale brown (Figs 11.11 and 11.12). The corals are slightly branched and have the appearance of very delicate shrubs. Rough material is fragile, but the gorgonin is usually removed in working, leaving the shorter, more stable calcareous sections. These sections are sometimes dyed – usually a rich red colour – when they are worked (Fig. 11.13).

Black and golden corals

These corals come predominantly from the Malaysian archipelago, the Red Sea, the Mediterranean and New Zealand. There are several types of black coral used for jewellery. They have skeletons made of a horny material instead of calcium carbonate.

It can be extremely difficult to tell the various black corals apart, even in their rough form, unless the whole coral is present. When made into jewellery or carvings, the task is often impossible.

Figure 11.4 *Heliopora coerulea*, 'blue coral', rough.

Figure 11.5 Blue coral necklace.

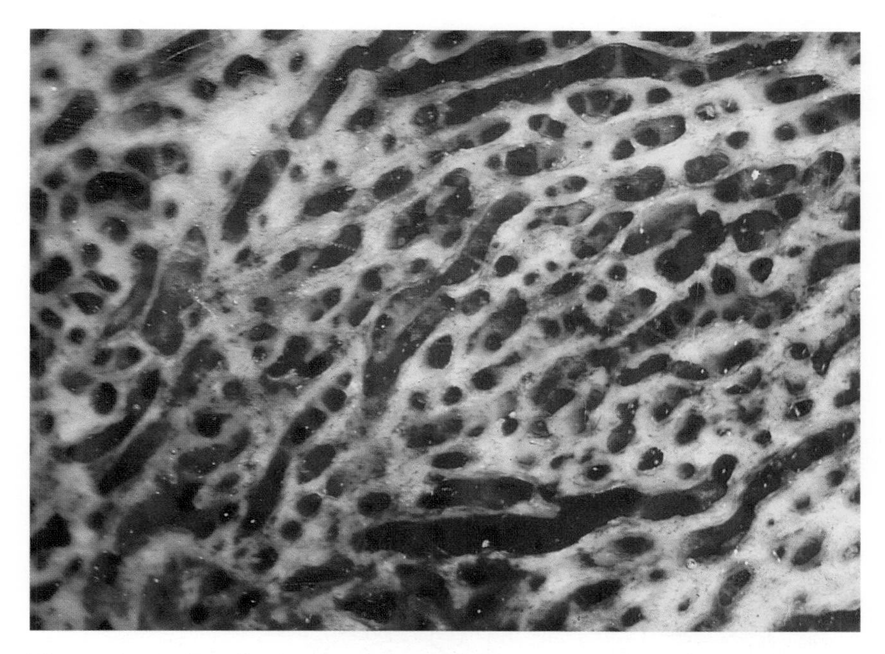

Figure 11.6 Detail of colour enhanced blue coral, showing accumulation of dye.

Figure 11.7 Triplet structure of section of blue coral necklace.

Figure 11.8 *Melithaea ocracea*, 'red soft coral', rough.

Most of the black corals used belong to the order Antipatharia (Fig. 11.14). They can be branched or unbranched, and they have concentric layers of horny material around a central canal, which may be very fine (the same applies to golden corals of the order Antipatharia (Figs 11.17 and 11.18)). Common to the antipatharians is that they all have spines. Most of these are minute and, in many cases, are only present

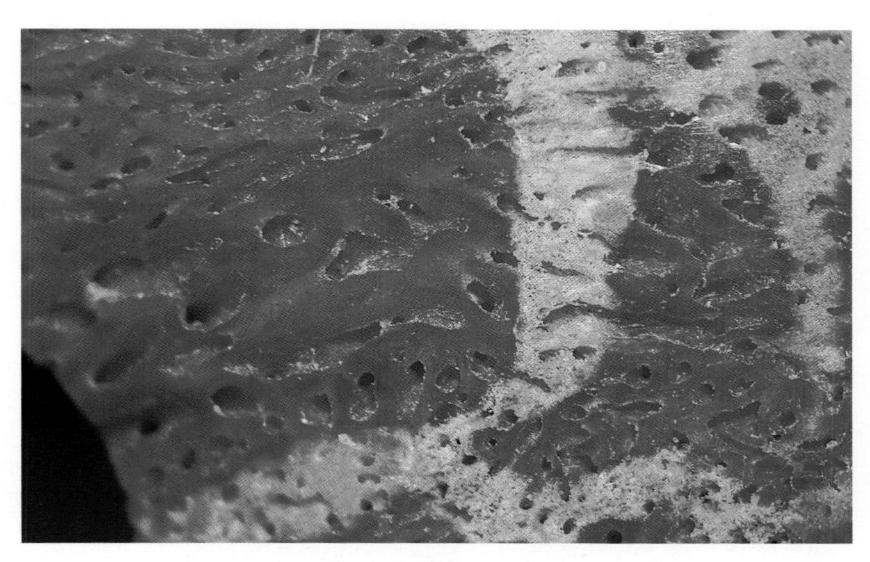

Figure 11.9 Detail of red soft coral, showing structure.

Figure 11.10 Red soft coral beads and disc.

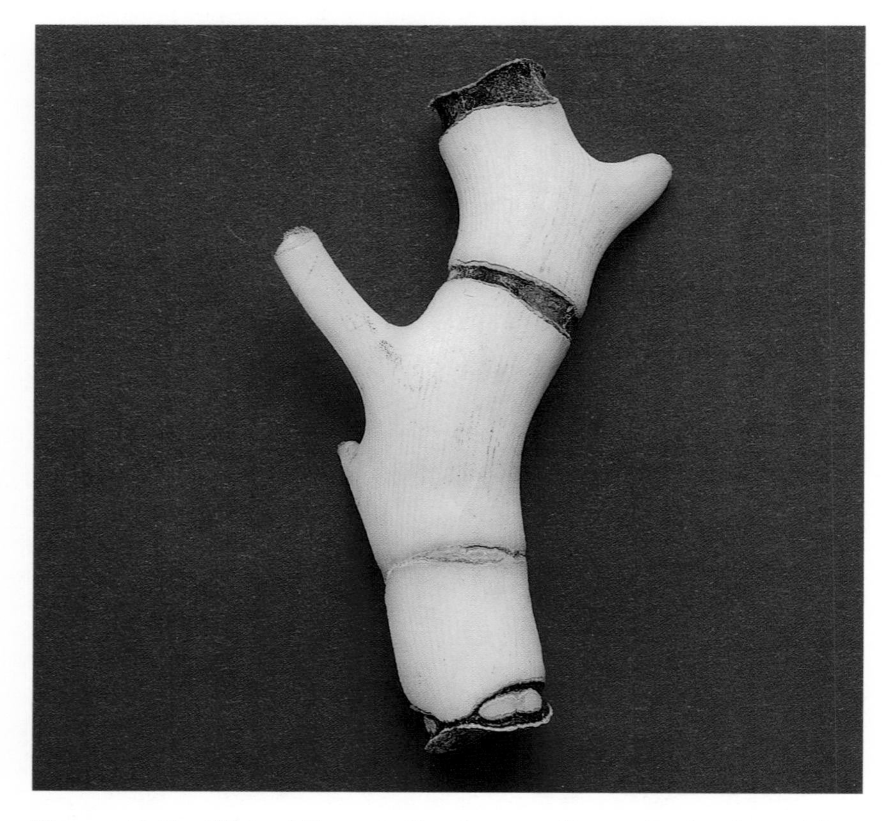

Figure 11.11 White isid coral, 'bamboo coral', rough, showing calcium carbonate sections interspersed with gorgonin.

on the finer branch tips, where they give an abrasive surface to the unpolished coral. On close inspection some of the spines appear almost translucent. The thicker stems may appear completely free of spines, which is one of the reasons that they are difficult to identify and may be confused with corals of other orders.

Some of the antipatharians have a surface that, when viewed under magnification, appears slightly bubbly (Fig. 11.15). This surface may remain when the coral is worked. *Stichopathes* and *Cirrhipathes* are examples of this. They are sometimes called 'wire corals' as they grow in a coiled unbranched form, which can easily be cut and made into circular bangles by joining the two ends. *Leiopathes* is another antipatharian, but it has a naturally high lustre, even before polishing. It often shows growth rings in a paler colour (Figs 11.14 and 11.16). It is well known around the Mediterranean.

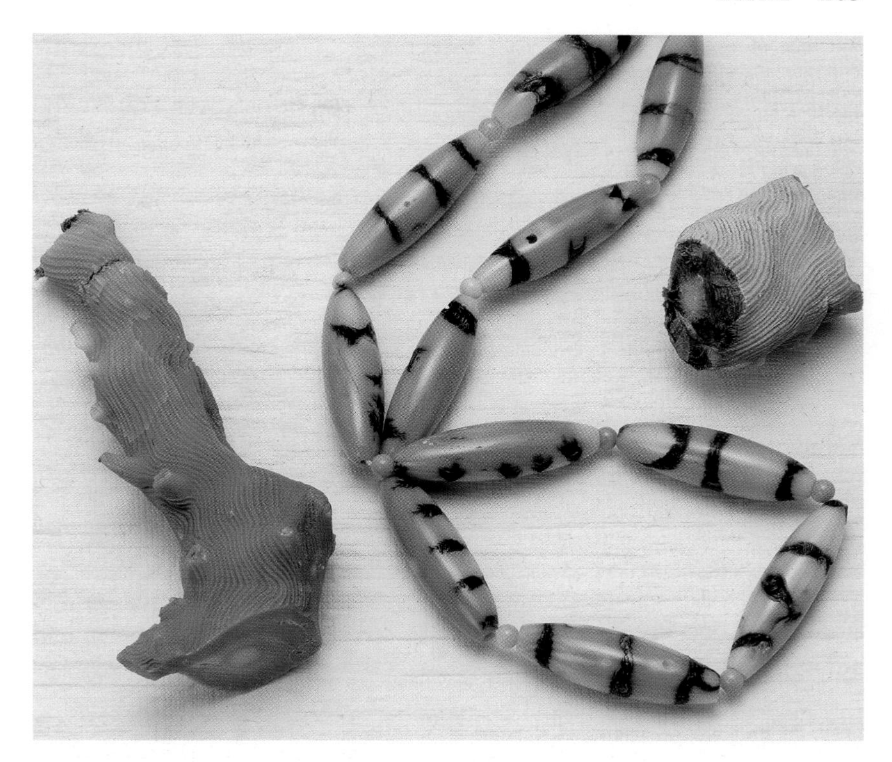

Figure 11.12 Brown bamboo coral: necklace and rough.

Figure 11.13 Detail of dyed bamboo coral, showing accumulation of dye.

Another Mediterranean black coral is *Gerardia*, of the order Zoanthiniaria. It is brownish to black, with a smooth surface, and can grow to 2 metres tall with a very sturdy 'trunk'. It has been known and used for centuries.

Gorgonian black corals of the order Alcyonacea are also used for gem purposes. They have no central canal, and their growth rings may be more irregular than those of the antipatharians. Their texture is also more fibrous and never displays any translucency. Gorgonians never have spines, but may have fine grooves (Fig. 11.19).

Both antipatharian and gorgonian corals can also occur in golden varieties. Here there is even more difficulty in identifying the species with absolute accuracy, as black coral can be bleached to a golden colour. The bleach can permeate quite deep into the coral, and it can be impossible to tell – without complicated tests – whether the golden colour is natural or bleached (Figs 11.17 and 11.18).

Fossilised corals

Fossilised corals used today are of the orders Scleractinia, Rugosa and Tabulata. The latter two are extinct. All three types of coral had

Figure 11.14 Black antipatharian coral rough and beads. Top and beads: *Leiopathes*.

Figure 11.15 Surface of black antipatharian coral rough (magnified).

Figure 11.16 Black *Leiopathes* bead, showing growth rings in lighter colour.

Figure 11.17 Detail of golden antipatharian coral bead detail, showing central canal. Possibly bleached.

Figure 11.18 Golden antipatharian coral necklace. Possibly bleached black coral.

Figure 11.19 Slice of golden gorgonian coral, showing growth rings and absence of central canal. Natural colour (magnified).

Figure 11.20 Fossilised coral beads and cabochons.

calcareous skeletons which have, in most cases, become embedded in sediment. With time, they have been impregnated with, or altered into, various minerals. Sometimes they become silicified, in which case they are much harder than gem coral.

All living scleratinians have completely white skeletons, yet fossilised corals come in a variety of soft brown and pink- or yellow-brown colours. These colours are caused by geological processes, after the coral has died (Fig. 11.20).

Fossil corals preserved in limestone are often seen in polished 'marbles' used for ornamental or decorative purposes, but being relatively soft this material is less suitable for jewellery. Silicified fossil corals, being harder, are sometimes seen in jewellery. This type of material mostly comes from Antigua, India and China. Some fossilised material has earlier been found in Wiltshire, England, but this is now only seen in items such as old jewellery.

Treatments and uses

Being diverse in structure, the various gem corals are treated in different ways.

- *Corallium* corals can be cut and tumbled to produce beads.
- All corals can be carved. *Corallium* corals take the most intricate carving (Fig. 11.25), while those consisting of a horny material cannot be carved with such fine detail. Blue coral and red soft coral are also too brittle and of too coarse texture to take fine and intricate carving.
- *Corallium* corals, the calcareous part of bamboo corals, and black and golden corals take a good polish (Figs 11.1, 11.12, 11.14 and 11.18).
- Blue coral is matt and chalky looking, and can only be polished if it is first impregnated with a stabilising resin. The result is a sheen on the surface, rather than a high polish (Fig. 11.5).
- Red soft coral is also matt, but has a rough surface. It must be impregnated with resin in order to be comfortable when worn. The result is, like blue coral, a sheen rather than a high polish (Fig. 11.9).
- The resin used to impregnate blue coral and red soft coral sometimes contains a pigment to enhance or darken the natural colour of the coral (Fig. 11.6).
- Thin sections of blue coral and red soft coral can be capped and backed with plastic to give a smooth, shiny surface (Fig. 11.7).
- The main body of a coral skeleton of *Corallium*, which can be very thick, can be carved into statues and figures of a substantial size –

up to 30 centimetres tall, or more. The finer branches of the coral skeleton are usually used to make beads (Figs 11.1 and 11.25).

- Pieces of solid, calcareous coral can be dyed. This is frequently seen in large, inexpensive pieces, often coloured bright red. The dye is only on the surface, but it is very difficult to tell exactly which species has been used as the raw material. The dye is applied after the coral has been worked (Fig. 11.13).
- Small pieces of *Corallium* coral can also be dyed. The colour used is more convincing than with the cheaper coral, and the dye is intended to deceive.
- Most *Corallium* coral is bleached before being worked, to clean off any impurities. This is also supposed to bring out the colour of the material. However, if it is overdone, the surface can be damaged, and the high lustre will in time deteriorate to a dull, opaque finish.
- Blue coral and red soft coral can be capped with plastic to stabilise them. They can also be backed with plastic to give more bulk (Fig. 11.7).
- Black corals that consist of a horny material can be bleached with hydrogen peroxide – in much the same way as can human hair, which also consists of a horny material. The resulting coral resembles golden corals, which are more rare than black. The effect does not penetrate to the core, but it is very convincing, and thus difficult to identify without cutting into the material.
- Research is ongoing into the possibilities of using some types of corals for bone grafts, and as artificial eyeballs, as the human body is less likely to reject coral than man-made substances.

SIMULANTS

- **Shell** is the most difficult simulant of coral to identify. In small pieces of polished material, for example cabochons, pink or white shell can look remarkably like coral. Each has a striated structure, but under high magnification, shell displays very much finer striations.
- **Glass** or **porcelain** are other simulants that can be convincing in small pieces. They are cold to the touch, and their glassy feel is similar to that of coral. They lack structure, and may show swirls of colour or air bubbles (Fig. 11.21).
- **Plastic** has been used to imitate coral since the early plastics came into production. It is not cold to the touch, and is lighter in weight. Like glass, plastic can show swirls of colour, air bubbles, and a complete lack of structure (Figs 11.22 and 11.23). Plastic is also softer than coral and its surface is easily scratched and damaged.
- **Reconstituted coral** – sometimes called Gilson coral, after the

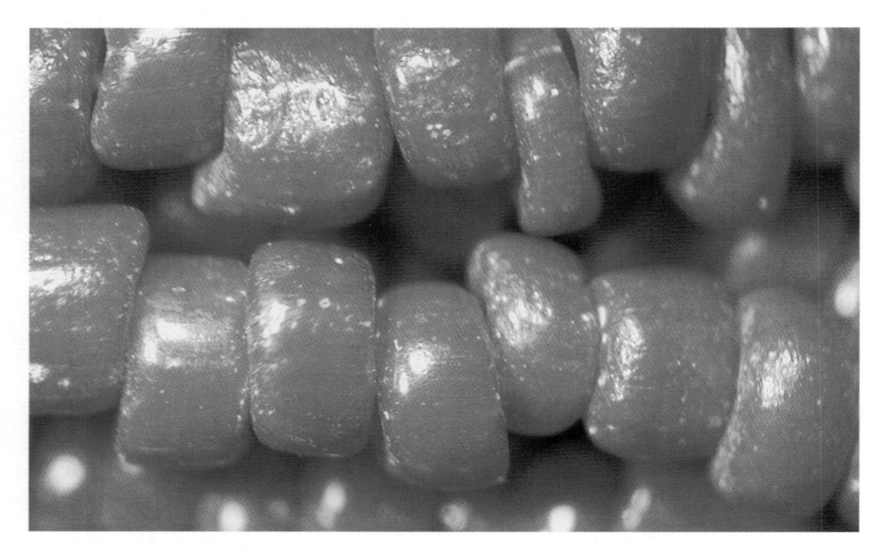

Figure 11.21 Glass beads imitating *C. rubrum* (magnified).

company that makes it – is made of crushed calcium carbonate with other materials added. It has been produced in colours from white to deep red – so-called ox blood colour. It has no surface pattern or structure, but can display pale, angular particles.

- **Jet** would not be used to imitate black coral, but, when polished, the two could be confused. No structure will be visible on the polished surface of jet, but growth rings and other structures may be visible in coral.

Tests and identification

As with all other organic materials, identification of coral is best done visually. It is most difficult to distinguish coral and shell from each other, and any chemical or hardness tests that can be carried out are likely give the same results for both materials.

Visual examination

Corallium corals

- Unpolished *Corallium* corals have a dull lustre and a finely ridged structure on the surface. The ridges run more or less longitudinally along the branches (Figs 11.2 and 11.3).

Figure 11.22 Celluloid brooch, imitating coral (magnified).

Figure 11.23 Plastic cabochon, imitating *C. rubrum*, showing swirls of colour.

- In polished *Corallium* corals the longitudinal ridges, though removed, are represented by a parallel striated pattern, with spacing of 0.25 to 0.5 millimetres between the lines. The striations are easily visible in red corals, but are harder to discern in pink or white varieties (Fig. 11.24).

Figure 11.24 *C. rubrum* beads, showing striations (magnified).

- In cross-section, these striations appear as radiating lines with very faint, concentric lines joining them, in a pattern somewhat resembling a spider's web.
- Lack of these radial lines in pink or white material may indicate that it is shell.
- Under high magnification, shell displays very fine striations, which are very much finer than those of the *Corallium* corals, though these can be difficult to see.
- Glass beads with rows of minute air bubbles can give the impression of coral structure, but their composition can be discerned when viewed under high magnification (Fig. 11.21).
- It can be extremely difficult to see if coral has been dyed, especially when it is mounted in a piece of jewellery. The width of the striations may be some indication of whether or not an item is made of *Corallium* coral (Fig. 11.13).
- Swirls of colour in an item that is cold to the touch indicate a glass simulant, though some shells are also variegated in colour.
- Complete lack of structure in an item that is cold to the touch indicates a glass simulant, or reconstituted coral material.
- Complete lack of structure in an item that is warm to the touch indicates a plastic simulant (Fig. 11.22).
- Air bubbles, mould marks, or swirls of colour in a material that is warm to the touch indicates a plastic imitation (Fig. 11.23).

Blue coral and red soft coral

- Blue coral and red soft coral are relatively easy to distinguish from other materials. They are not as yet copied, and they both have a very distinctive porous structure (Figs 11.6 and 11.9).
- Under magnification, resin-impregnated examples will show accumulations of this material in and around the tiny surface holes (Fig. 11.6).
- Impregnated coral will have a satin sheen instead of being dull and chalky (Fig. 11.10).
- Impregnated coral will feel smooth to the touch, instead of rough.
- The impregnating resin will display a total lack of natural structure (though air bubbles might be present). However, this would only be visible if a relatively large void is filled with resin.
- Dyed, impregnated coral may show accumulations of colour around the tiny holes, and in any cracks that may be present (Figs 11.6 and 11.7).
- Viewed from the side, coral doublets will show a clear, plastic cap (Fig. 11.7).

Bamboo corals

- Bamboo corals can be identified by the occasional, dark brown, gorgonin nodes that regularly intersperse the calcareous material, dividing the branches into sections (Figs 11.11 and 11.12).
- As with precious corals, unpolished bamboo corals have a parallel longitudinal ridged pattern on the surface of the calcareous sections. At about 1 millimetre apart, they are much coarser than those of precious corals (Figs 11.11 and 11.12).
- When polished, the surface ridges are removed but are represented by striations.
- Dyed coral may show streaks, patches, or accumulations of dye around any holes or cracks. The dye is only on the surface, so an area that has been chipped or cracked after the material was dyed may show a lack of colour (Fig. 11.13).

Black and golden corals

It is extremely difficult to distinguish between the various species of black and golden corals (see under 'Species'). The following therefore concentrates on the corals and their simulants.

- Black and golden corals, being made of proteins and not of calcium carbonate, are warm to the touch. Although slightly flexible when alive, the skeletons feel rigid.
- Black coral can take a high polish, but in some species it retains a natural sheen, and, viewed through a 10× lens, displays a structure of tiny bubble-like bumps on the surface.

- Polished black coral can be confused with polished jet. The latter shows no structure, while signs of growth rings may be visible in polished coral (Fig. 11.16). However, breaks or chips in the material may help with identification. Black coral's fracture is splintery, while that of jet is conchoidal.
- Total lack of structure in a specimen that is warm to the touch may indicate plastic.
- Mould marks, air bubbles or swirls of colour in a specimen that is warm to the touch indicate plastic.
- A rubbed or abraded area that looks dark, in an item apparently made of golden coral, indicates bleached black coral. However, this material is difficult to identify, as the bleaching usually takes place after working, so even the drill hole in a bead will be bleached (Figs 11.17 and 11.18).

Fossilised coral

- Fossilised corals lack a pattern of striations, but usually display small circles with radiating lines. Originally these circles would have been holes in the corals, but they are now infilled, and often silicified (Fig. 11.20).

Tests

There are gemmological tests that can be carried out on coral, but these are rarely feasible. Most other tests are inconclusive or destructive.

- **Ultraviolet light.** Black and golden coral are inert under UV light, as are plastics, which are the most likely simulant. Calcareous corals fluoresce under UV light in the same way as their most convincing simulants, shell. They both display a pale, chalky colour, and stronger fluorescence is noted in the paler material. Dark coral shows no discernible fluorescence.
- **Acid.** A drop of acid will cause calcareous corals to effervesce. This is destructive as it damages the surface of the material. Further, it is not a useful test as shell will react the same way.

Conservation status and availability

Living corals face many hazards. They are susceptible to climate change, as the algae with which many of them live are dependent upon an even water temperature. Corals can be affected by earthquakes or violent storms, which can uproot or break them, or bury them in sand. Fungal or bacterial infections also take their toll, and various creatures feed on corals, especially parrot fishes and crown-of-thorn starfish.

Humans have not helped the situation. Snorkelers and divers can damage corals, even while admiring them. Ships' anchors, as they drag on the sea floor, are another danger for the animals, as is any form of water pollution, including suntan oil.

Early fishing methods were also very destructive. Dynamite used to blow fishes to the surface also blew up the corals, and cyanide used in fishing induced a coma in the fishes being caught, making them easier to catch, but it killed the corals. Netting fishes inevitably also caught corals. Today explosives are still occasionally used, though this is now illegal.

Many gem corals have been overfished, with the result that there is now little left in some areas. An example of this is the Mediterranean. As most corals grow very slowly, it will take a long time for them to re-establish themselves. Some experimentation is under way with coral farming – for instance, in Japan and Hawaii pink *Corallium* coral is now being bred – but the long-term success of this has yet to be established.

The rules governing the conservation of the various corals are constantly changing, and many are now protected species and may not be collected at all. In some areas there is a quota for collection. At the time of writing, blue coral, *Heliopora coerulea*, and some black corals are listed on CITES Appendix II, which means that their trade is covered by very strict controls. Sadly, the controls are not always effective, and there is also a certain amount of poaching. Further, for many species of corals the controls are not nearly stringent enough and whole areas are legally harvested until they are bare.

It is advisable not to buy or import corals without seeking advice from an appropriate authority, and never to collect corals oneself.

PAST AND PRESENT USES

According to Greek mythology, there were three gorgons. They were sisters, and they all had wings, claws, enormous teeth, and snakes for hair. Medusa was the only sister who was mortal, and she was beheaded by Perseus in the course of one of his heroic feats. When he put the head down on a bed of leaves and seaweed, the snakes immediately turned to red coral. The gorgonian corals take their name from the myth.

All through the ages coral seems to have been as much prized for its supposed talismanic properties as for its beauty. It is only since the middle of the eighteenth century that it has been accepted that corals are animals and not plants. An Italian alchemist had suggested the possibility in the seventeenth century, but the idea had been dismissed as crazy. Even today, some people are not aware that corals are animals. This is probably because living corals look like plants, and many have shapes similar to small trees or shrubs. However, whether it was believed to be a plant or an animal, mankind's love of red coral goes back several thousand years.

Fragments of red coral amulets have been found in Neolithic graves in Switzerland, dating from about 8000 BC, and the oldest intact, coral object is a small idol, found in a Neolithic grave in Italy.

Coral is thought to have been used for barter in Asia about 5000 years ago. The material was favoured by the Celts, who lived in parts of Europe before the Roman invasion. The Celts used coral for decoration on shields and helmets, but the Romans used it mostly for trade or for its amuletic powers. These included the ability to still tempests, to ward off evils, to protect crops from blight, and to cure various diseases and enchantments.

During the Han dynasty (about 100 BC), the Chinese were so taken by coral, that they sent an expedition to the Mediterranean to investigate the possibilities of harvesting the material themselves. It was worn by mandarins of the second rank, and woven into, or embroidered onto, their robes. Surprisingly, one ancient culture that had access to coral did not make much use of it. The Egyptians preferred other red gems such as carnelians or garnets.

Black coral was also traded 2000 years ago, though not to the same extent. The Greeks believed that black coral could guard against scorpion stings, while the Romans thought it a potent love potion. In other parts of the world it was believed to cure rheumatism, ensure the healthy growth of children, and protect adults against sorcery.

It is known that black coral was being made into bracelets by the people of the Malay Archipelago in the mid-eighteenth century. They soaked it in coconut oil and heated it gently to form it, and sometimes inlaid it with gold for use by their chiefs. This treatment would have been possible because, being a horny material, it would be to some degree thermoplastic.

In Europe in the Middle Ages, red coral was given to babies to chew on, in order to 'fasten and strengthen' their teeth. An added benefit was that coral was believed to ward off evil spirits. In powdered form it was fed to babies in milk to prevent fits. In the nineteenth and early twentieth centuries it was still being given to babies, as a traditional Christening present, for example as a necklace, or a teether attached to a rattle.

Some superstitions were based on the belief that coral could change colour. It was thought that, when worn by men, coral was red; worn by women it was paler; and on the sick it was very pale. A variation of this belief was that the depth of colour was an indication of the health of the wearer, with deep, red coral signifying excellent health, and the palest shades signifying very poor health.

Figure 11.25 'Karon's boat'. *Corallium* coral, silver and ebony. Possibly Italian, about 1650. *Rosenborg Castle, Copenhagen, Denmark.*

Although coral was used all through the ages (largely for its talismanic properties), there was a period following the fall of the Roman Empire during which little was used in Europe. In the thirteenth century, coral was again being used, mostly for rosaries in Mediterranean countries. The coral industry really began to develop 100 years later. Many towns were originally involved, with centres in Genoa, Trapani, and Naples in Italy, and Marseilles in France.

By the seventeenth century, coral was being fashioned into a variety of items, such as combs, candlesticks, figures and statues, jugs, boxes, and jewellery. Some objects were mounted in silver or silver gilt, and some jewellery was mounted in gold. As with most of the other organic gem and ornamental materials, coral found great favour at the European courts, and beautiful carvings can today be found in royal collections and museums (Fig. 11.25).

In the nineteenth century the work had decreased, and a town near Naples, called Torre del Greco, had taken over as the main centre for coral carving. It remains the European centre today, and the most famous worldwide, though there is now also a large coral industry in the Far East.

The African countries bordering the western end of the Mediterranean have fished for coral for a thousand years. It has been used not only as decoration, but also for its talismanic properties, to protect both children, and adults. Even houses have been adorned with branches of coral to guard against storms and other hazards. Benin (the former African kingdom which is now part of Nigeria), has for centuries been renowned for its high quality of craftsmanship using materials such as ivory and metals. Coral has also been used there for several hundred years. It first appeared as a result of trade with the Portuguese in the sixteenth century. The local name for it, 'Eshugu', means wealth and rank. Vast quantities of small beads have been worn in ceremonies by the ruler, or 'Oba'. In Nigeria coral has always been, and to some extent remains, a status symbol. Today the preference is for large beads.

As with most organics, coral was not much used in the United States. Relatively recently, the Native Americans began to use red coral, often combined with turquoise beads, and possibly with feathers.

Coral has been fished in the Mediterranean for thousands of years, but the main coral finds in the Far East have occurred in the last 50 to 100 years. The Japanese first found *Corallium* corals in the seas around their islands in 1830, and to begin with they sold it all to Italy. More coral was found off the coast of China in the 1920s, but it was in the 1960s to 1980s that large quantities were found in various parts of the Pacific.

Today coral is mostly used for jewellery. A little is carved into figures, and some is used for medical purposes and research.

12 Miscellaneous organics

The list of organics that have been used as some form of gem or ornamental material is almost unlimited. The following is not intended as an exhaustive list, but as a brief introduction to a few of the less well-known organics.

Baleen

Some of the world's largest whales, including the largest of all mammals, the blue whale, do not develop teeth but instead have hundreds of overlapping plates of baleen hanging like curtains from their upper jaws.

These plates are made up of three types of keratin – the organic protein that makes up hair, horn, nails, and so forth. It forms as: rows of fine tubes; a cementing matrix to hold them together; and a thin coating layer on either side. The keratin tubes become slightly mineralised which gives them added strength.

Although only a few millimetres thick, in some whales the baleen plates can grow up to 4 metres long. They grow continuously, as do human fingernails, but the cementing and covering layers are worn away with use, resulting in solid plates with fringed bristly edges. Whale baleen comes in varying colours: off-white, brown and almost black, depending on the species (Fig. 12.1).

In spite of their size these large whales live mostly on tiny, shrimp-like creatures called krill, eating them by taking in huge mouthfuls of water and then siphoning the water out again through the baleen fringes, so that the krill get caught as though in a sieve.

Baleen is sometimes also (misleadingly) called whalebone. It is a tough material and is thermoplastic. It was popular for use as bristle in heavy duty brushes, and for corset stays and umbrella ribs as it was slightly flexible. It was also engraved and decorated as scrimshaw – the whalers' craft (see Chapter 3, 'Ivory'). The solid part of a baleen plate looks much like horn and has been put to many of the same uses, for

Figure 12.1
Baleen plate.

example in the manufacture of small boxes. It can be difficult, if not impossible, to tell the two materials apart in a finished piece, as heating and moulding destroys much of the structure of the material. Tests, such as burning a tiny scraping, will not distinguish between them, as they will both give off the same smell (see Chapter 7, 'Horn').

Byssus

Byssus is the name given to the silky threads produced by some bivalve molluscs.

The threads are secreted as a liquid by a gland on the animal's foot. The liquid hardens on contact with water. The threads are used to attach the mollusc to a firm surface on the seabed. If the animal wishes to move on, it cuts itself loose by breaking the threads, and reattaches itself elsewhere by growing some more threads. Some molluscs use their byssus as a defence mechanism against predators, which they tie up.

Byssal thread has been used for many centuries as a very exclusive yarn. The threads of the noble pen shell were the most used as they were a beautiful, iridescent, golden-bronze colour and could reach 60 centimetres long. Hundreds of shells went into the production of, for example, one pair of gloves, as byssal thread is very fine and light.

Cloth made of byssal thread was known to various ancient cultures. It is thought that it was used by the pharaohs in Egypt, and that the 'golden fleece' in Greek mythology probably refers to byssal cloth. There are only sparse references to the material over the centuries, but it is known that it was used in some way by the Romans, and later by wealthy people around the Mediterranean for stockings and gloves. It has always been regarded as the ultimate luxury.

There is a fine example of a pair of gloves made of byssal threads in the Natural History Museum in London, where they are on display in the Invertebrate Galleries.

CALABASH

Calabashes are gourdes. The ripe fruit are heavy and soft, but get lighter as they dry and their skins harden. They turn brown when they are dried.

Calabashes grow in tropical climates, and were used by local populations mainly as containers for storage and for carrying water. Medicinally, they were consumed as a cure for colds and baldness, and it is said that pieces of shell were, on occasion, surgically inserted into a patient's broken skull as a protective plate. There is no record of the success rate of this application.

The Europeans brought calabashes back to Europe as curiosities. The first ones were carried home by early explorers and mounted as cups on silver or silver gilt stands, purely for decoration. Examples can be seen in some museums (see 'Coconut').

COCONUT

Coconut palms, *Cocos nucifera*, are extremely versatile plants, as every part of the tree can be put to use. They grow worldwide in warm climates, especially by the sea, and can reach 30 metres tall. The fruits have a smooth outer layer, inside which is a fibrous layer protecting the coconut. The rough, hard, brown shell of the nut contains white flesh and liquid.

The liquid is very nourishing, and the flesh is a rich source of protein that can be pressed to produce oil. The sap of the tree is fermented to produce an alcoholic drink, the trunks are used as timber, the leaves are plaited to make thatch, and the fibrous layer of the fruit is made into coir matting, rope or compost.

Figure 12.2 Section of belt decorated with shells, coconut shell discs and glass beads. Thailand.

Figure 12.3 Coconut wood salad servers.

The nut shells make useful utensils or can be used as a decorative material. When the rough, fibrous layer is ground away the resulting brown surface can take a good polish and is an attractive, warm brown colour. The material can be cut into any shape and used for inexpensive jewellery, as buttons, or as ornamentation on clothes (Figs 12.2 and 12.10). Left whole, the shells have also been made into decorative cups and mounted in silver or silver gilt, or on carved ivory stands, in the same way as calabashes or ostrich eggs (Fig. 12.12).

More recently the wood of the trunk has been fashioned into decorative items such as boxes or small dishes. The wood has a very distinctive, fibrous looking grain and is sometimes called 'porcupine wood' (Fig. 12.3).

FEATHERS

Over the millennia feathers have been one of the most utilised of the organic materials. They have been easy to come by, and in most cases have needed no preparation or treatment before they could be used.

Feathers are keratinous growths emerging from the skin of birds. For these creatures they have two main purposes: insulation and flight. A secondary purpose is for display, either to show off and frighten an enemy, or to show off and attract a mate.

For ages humans have used birds' feathers for insulation in clothing or bedding. Although some experiments were carried out, it proved impossible for humans to use birds' feathers for flight. However, they proved to be excellent for human display.

Initially feathers were regarded as a status symbol. They were very much part of any tribal attire in all parts of the world. In Africa feathered head-dresses or necklaces were worn by shamans or 'witch doctors', possibly along with animal teeth and bird beaks. In South America whole capes were covered in bright feathers, for example the scarlet feathers of the ibis. Native Americans wore them as head-dresses or jewellery, and, in some cases, accorded them talismanic powers. Even in Scotland, the number of eagles' feathers worn on the bonnet signifies rank, and although it is an old custom, the rule still applies. Only the clan chief may wear three eagle feathers, and a chieftain may wear two.

Feathers have always played a great part in ceremonial attire, from the plume in the helmet of a cavalryman to the feathers on the hat of a wedding guest. In some remote areas of the Far East, head-dresses made of the feathers of 20 or more exotic birds were worn by all male tribal members on every ceremonial occasion – a custom which caused some birds to become almost extinct.

In China there was a passion for kingfisher feathers. The iridescent turquoise feathers were used instead of paint or enamel, and mounted with great creativity on metal, wood or even papier maché, to make such items as screens, fans and jewellery. They were popular for head-dresses, especially for bridal head-dresses. These were extremely elaborate and often combined fine, feather-covered metalwork with coral and pearls – the pearls hanging in long fringes at the front to cover the face like a veil. Many of the kingfishers were imported from Cambodia, and the use of these feathers goes back about 1000 years. When viewed under magnification, the fine, ridged appearance of a feather's structure can be discerned.

Today feathers are again popular in jewellery, most commonly hanging in pendant ear-rings or necklaces. They are often dyed, as most of the birds with exotic, coloured plumage are protected by conservation treaties. The feathers used come mostly from domestic fowl such as ducks, or from game birds.

Fossilised molluscs

Many molluscs have a shell with a pearly surface lustre. This is due to optical reflection effects from calcium carbonate, crystallising in the

Figure 12.4 Two pieces of gem ammolite, on ammolite rough.

form of aragonite. During fossilisation the aragonite – which is unstable – is usually replaced by calcite, which very seldom has a pearly lustre.

Ammonites are extinct, marine molluscs, distantly related to nautilus. When cut in half and polished they often display attractive formations of crystals within the outline of the original shell. Some contain crystals of the brassy-coloured mineral pyrite. Ammonites occur worldwide and are regarded as collectors' items, or are used for ornamental purposes. In Yorkshire, England, they are found in the same area as jet, and are sometimes combined with that material and made into items such as chessboards.

A few molluscs retain the aragonite, and some of these display a wonderful iridescence. **Ammolite** is the name given to a gem material that derives from the fossilised shells of ammonites of the two species *Placenticeras meeki* and *P. intecalare*. These shells are found in southern Alberta, Canada, and are about 70 million years old. The shells measure up to 90 centimetres in diameter, and are preserved in grey shale. They display iridescent colours, predominantly green and red, though blues, yellows, oranges and purples can also be present. The surface rarely appears completely smooth, but usually has a cracked look, the coloured areas interspersed with dark brown matrix. It sometimes resembles a miniature, stained glass window (Fig. 12.4).

Some gem quality pieces of ammolite are stabilised with epoxy or other resin before cutting, to prevent the delicate surface from flaking. However, some material is cut without prior treatment. The cut pieces are then polished, and finished in one of three ways. The most valuable and rare pieces are solid shell, with no matrix on the back. Thinner slices are made into doublets to reinforce the material by adding an extra layer of natural matrix on the back. Triplet stones have both an extra matrix back, and a cap of synthetic spinel or quartz.

The material was not commercially mined until 1981. In the past it has been sold under the names of 'Calcentine' and 'Korite'. Prior to its discovery, ammolite was known to the Native Americans of the Blackfoot tribe, who believed that the gem had magical properties giving them supremacy over the buffalo.

Lumachella is a similar material of iridescent, multicoloured shell, though it is not as spectacular as ammolite. It tends to appear more as fragments in a matrix of limestone, giving an effect of marble with iridescent patches. It is found in Italy and Austria and is used as a decorative marble – sometimes in mosaics. It is not used for setting in jewellery.

HAIR

Hair jewellery was popular from the late eighteenth to the late nineteenth centuries. In early examples just a single lock of hair was mounted in a ring, brooch or locket, possibly fanned out to resemble feathers or positioned to look like a tree. Later a new vogue arose of working with hair, and the results were much more elaborate. The simplest was a plait or woven section behind a piece of rock crystal or glass, in a pendant or brooch. More elaborate versions had gold wire ornamentation – often depicting initials – in front of the hair, or a surround of gemstones or seed pearls. Some pieces contained the hair from each member of a whole family.

Much more intricate were the items made almost entirely of hair, with just a clasp or fixture of silver or gold. The hair in these delicate pieces could be woven into many shapes such as bows, loops and hollow spheres, and worn as brooches or pendant ear-rings. Another style was a bracelet or necklace made up of narrow ropes of woven hair.

Horsehair from the animal's mane or tail was sometimes used as it was much coarser than human hair, and therefore easier to work. However, horsehair was regarded as inferior. True hair jewellery had to be made from human hair, and was mostly worn as a symbol of love or affection for the owner of the hair.

Hair is a keratinous material, quite strong and totally flexible. Long hair was required for fashioning, for example a bracelet needed hair that was at least 60 centimetres long. Before it was worked the hair was boiled. It was then plaited or woven like lace around a mould or firm base, and boiled again to keep the shape.

Hair jewellery was acceptable as mourning jewellery, indeed the hair quite often came from a departed loved one, becoming a lasting link. In America it was in great favour during the Civil War. The fashion reached its peak in Europe in the mid-nineteenth century. By this time young ladies were making the jewellery themselves, at home, as they might otherwise have made lace. It was even possible to buy starter kits to learn the craft.

An entirely different form of hair jewellery is that made from the hairs from elephants' tails. As they are part of an elephant, the hair is nowadays covered by trade bans, but in the mid-1900s expanding bangles made of the material were very popular. The hair is black and very coarse, and was usually wound in a circle with decorative knots, possibly trimmed with gold. Today similar bangles may be bought, but coarse hair from other animals is used instead of elephant hair.

HORNBILL IVORY

Hornbill ivory comes from the helmeted hornbill, *Rhinoplax vigil.* One of the group of birds called hornbills, it is a native of central South East Asia. It is the emblem of West Kalimantan Province in Indonesia.

In Malaysia it is sometimes called the 'Kill your mother-in-law bird', due to an old legend connected with the raucous cackle it produces.

It carries an unusual casque rising vertically from the top of its bill, which has been coveted for 2000 years as a material for carving. The purpose of the casque is unknown, but it is probably used as a tool to tamp the earth and leaves which are used to seal up the nest (with the female inside), while the eggs are hatching.

Only the front 2 centimetres or so are made of solid material and suitable for carving. The rest of the casque is a fibrous structure, like the inside of the bill itself. The bill is yellow but the casque has a red coating on the top and sides, extending over part of the bill.

The casque is chemically different from ivory, so the name 'hornbill ivory' is misleading. It is made of a keratinous material that can be carved, turned, or, being thermoplastic, flattened or moulded. The red surface may be totally or partially removed in working. With time the yellow colour fades to a dark, creamy hue.

The material was much loved by the Chinese who carved it into snuff bottles, belt buckles, jewellery and other small items, some of which were exported to Europe. The Japanese carved it into small items such as netsuke.

The hornbill has been overhunted and is endangered. Articles made from its bill and casque are rarely seen, and if seen, are probably not recognised.

INSECT WINGS

The wings of some insects have been used for decorative purposes, either as jewellery or as ornamentation.

In China, butterfly wings were occasionally used in the same way as kingfisher feathers. Small sections of the brightly coloured wings were cut out and glued onto a background, possibly silver, the effect somewhat resembling enamel.

Scarab beetles, *Scarabaeus sacer*, also known as dung rollers, are indigenous to Egypt, where they were regarded as sacred in ancient times, and whole beetles were mummified. This was still being done as recently as the 1950s, and the beetles sold as jewellery. Today they are copied in carved stone.

Worldwide there are many species of beetle that have brightly coloured, iridescent wing cases that can be made into jewellery, often as necklaces or pendant ear-rings. In Western Europe the blue-green wing cases of the tansy beetle, *Chrysolina graminis*, were popular in the nineteenth century. They were used in jewellery, and were also sewn – like sequins – onto purses, shawls and dresses.

Figure 12.5 Chinese lacquer brooch (magnified).

Today insect wings are seldom used, though they are occasionally incorporated into jewellery.

Lacquer

The lacquer used in traditional Chinese or Japanese carved lacquer ware is derived from the resin of the tree *Rhus vernicifera*, which grows wild throughout China, especially on high plateaux.

The resin is extracted from the tree in much the same way as rubber. A cut is made in the tree trunk and the transparent resin pours into a little cup tied beneath the cut. As it comes into contact with the air, the resin becomes opaque and thickens. Before use it is boiled and filtered, and mineral pigments are added to give colour. Examples are: cinnabar for bright red, or iron oxide for dark red.

Today it often has oil added and is moulded before being carved, but in olden days it was used in its pure state and painted onto a piece of hemp cloth, with up to 300 thin layers applied, to obtain sufficient thickness for carving. The multiple layers give an added brilliance and toughness to the end product. The hemp cloth may be visible as a slightly hairy backing to a piece of lacquer work.

Figure 12.6 Shagreen and elephant ivory games pieces. England. 1930s.

Wood can also be lacquered, which preserves and protects it. The lacquer surface can further be painted, or inlaid, for example with mother-of-pearl or metal foil.

Lacquer can be used on such varied items as furniture and small brooches (Fig. 12.5). Boxes have always been very popular.

SHAGREEN

Shagreen is a leather made from the skin of fish of the ray family, especially stingray. It can also be made from sharkskin.

The natural skin is covered with denticles or bumps called placoid scales, which are unique to sharks, rays and skates. These consist largely of calcium carbonate and calcium phosphate and are more mineralised than normal fish scales (which have a higher collagen content). The denticles appear as spots, some tiny, and some larger and spiky.

Traditionally the skin is treated by a process involving being scraped, stretched, dried, filed smooth, dyed and varnished, and finally polished. In this form it is a rawhide and is not tanned. The most common colour for dying has been green, but it can take any colour. If it is not

Figure 12.7 Vegetable ivory kernels: natural, polished and broken.

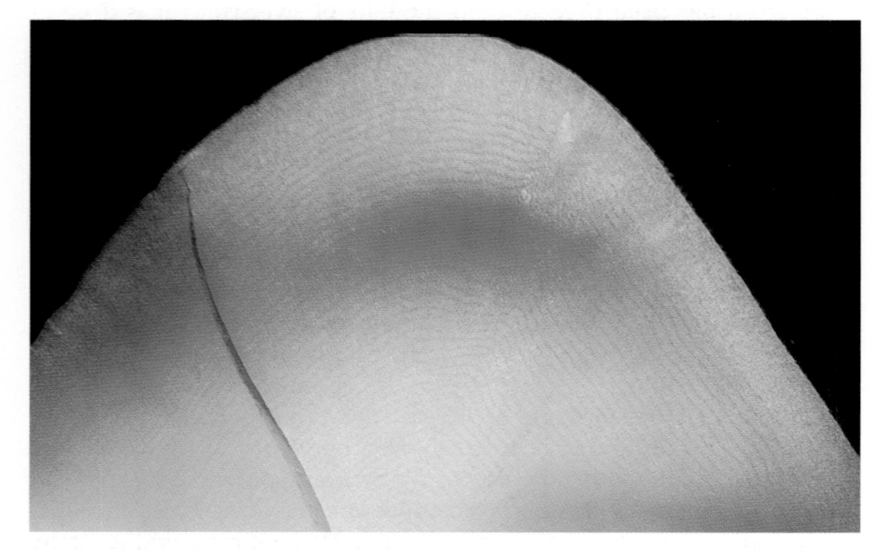

Figure 12.8 Vegetable ivory structure.

Figure 12.9 Vegetable ivory carvings: sewing needle case, hippo, and buttons.

filed until after it has been dyed and varnished the result is a coloured background with white spots; however, an all-over colour is more usual. In the 1920s a method was found whereby the skin could be decalcified and tanned, which softened it.

Shagreen has been used to bind the hilts of swords, as it is decorative and gives a good grip. Japanese warriors had discovered this by the eighth century, and the material was again put to this use in the American Civil War.

In Asia, in the fifteenth century, it was used to cover trunks and coffers. Three hundred years later in Europe it was occasionally used as an inlay or veneer for luxury items. Later, in the art deco movement (the decorative style of the 1920s and 1930s), it became very popular and was used to decorate all manner of objects, from jewellery to furniture.

Boxes, humidifiers, toiletry sets and games pieces were the more usual bases for the material, often in conjunction with ivory edging (Fig. 12.6). Today shagreen is still manufactured, though on a much smaller scale. Popular items are picture frames and small boxes. In rare instances it also appears on jewellery.

It is uncommon to find a shagreen fake, but some were made at the height of the material's popularity in the twentieth century. They are not convincing as the natural product has a slightly uneven surface due to the denticles, while a fake is made of a printed laminate and is totally smooth.

Other fish

Fish leather is not confined to shagreen. Theoretically any fish skin can be turned into leather, and it is often stronger than its mammalian equivalent. Fish leather has been made for centuries and has proved a useful alternative in times of need, such as during the Second World War, when it was made into shoes.

The process is different to that used for shagreen. The skins must be descaled, delimed and preserved, in order to remove oils and the fishy smell. They are then tanned and dyed, before being stretched and polished. Today fish leather is becoming very fashionable, with salmon skin leading the way.

Vegetable ivory

Vegetable ivory, as its name suggests, is of vegetable origin. It is derived from the nuts of several plant species, the two most common being the doum palm, *Hyphaene thebaica*, from Africa, and the tagua, *Phytelephas macrocarpa*, from South America. Other species producing similar nuts are found in South East Asia.

The form and structure of the nut varies with each species. The doum palm produces fruits the size of small oranges, that each hold one kernel (nut), while the tagua produces fruits the size of human heads, that contain multiple kernels. Both types of kernel are about the size of a chicken's egg. The tagua is the nut most used today.

When it is ripe, the fruit of the tagua palm splits, and the kernels drop out. The nuts are still soft inside and are edible. They must be carefully dried and possibly fumigated to ensure that they are insect free before they can be worked. They have a tendency to produce a star-shaped crack in the centre as they dry.

The dried nut is very hard, with a dark brown husk. The interior is a very slightly translucent creamy white (Fig. 12.7). In cross-section the plant cells of the nut, which consists of cellulose, show as fine, regularly spaced, concentric lines (Fig. 12.8). Viewed from any angle under magnification the nut displays a fine, regular, densely packed spotted appearance. This is diagnostic for vegetable ivory. In finished, dyed pieces, a pretty, moiré silk type of pattern may be visible.

In the unlikely event that it is impossible to determine whether an item is vegetable ivory, a drop of sulphuric acid will cause a pink spot to appear on cellulose after some minutes. However, this is irreversible and the spot cannot be polished away as the acid seeps into the material. It is therefore a destructive test, and not recommended.

Vegetable ivory can be carved, turned or engraved, and it stains very readily. Due to its small size its uses have been limited, though several

Figure 12.10 Necklace of coconut shell, ostrich egg shell, snake and fish vertebrae. *Jewellery by kind permission of Stefany Tomalin.*

pieces can be combined. Good examples of this are the sewing requisites so popular in Victorian times, such as thimble holders and needle cases. These were turned on a lathe in sections, and the sections were then screwed together (Fig. 12.9).

In the latter half of the nineteenth century a whole industry grew up around the manufacture of vegetable ivory buttons. Some were intricately carved and multicoloured (Fig. 12.9), while others were completely plain. With the advent of early plastics, this industry faded. There was a lot of wastage in the production of buttons, and it is said

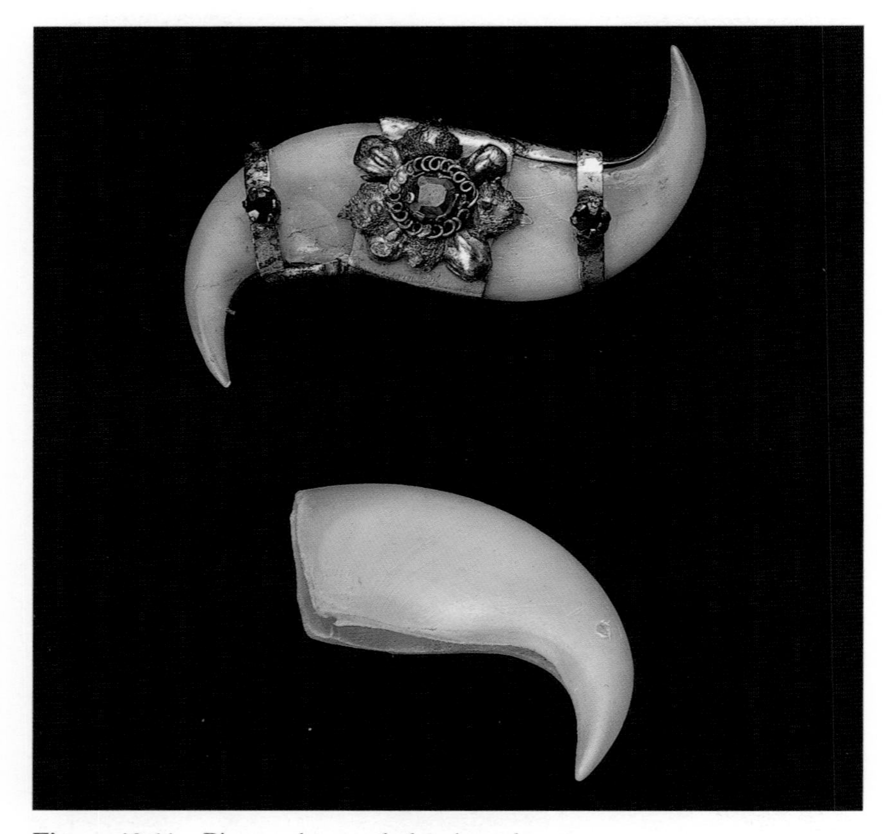

Figure 12.11 Big cat claw, and claw brooch.

that the resulting powder was sent to munitions factories for use in the manufacture of explosives.

In the absence of animal ivory, vegetable ivory is again popular for small carvings, many of which reflect the shape of the nut (Fig. 12.9). The outer, brown husk is sometimes left for added effect. Today there is a large production of vegetable ivory netsuke in the Far East, and the nut is also used for scrimshaw. A small amount of buttons is still produced.

Other miscellaneous organics

Seeds and nuts are easy to find. They are not durable, so their early uses can only be guessed at, but it seems obvious that they would have been threaded in some way, and made into pretty necklaces. They can also be carved to make small, decorative objects.

Figure 12.12 Two decorative cups of ostrich egg shell and ivory. Lorenz Spengler and J.E. Banert. 1757. *Rosenborg Museum, Copenhagen, Denmark.*

Small bones, for example poultry bones and fish or snake vertebrae, can similarly be threaded, as can sharks' teeth (Fig. 12.10). These teeth can sometimes be found washed up on beaches, in areas of the world inhabited by the sharks. They can also be purchased in fossilised form, though these are sold as curiosities and collectors' items.

Animal claws are made of keratin, and have been used for adornment for thousands of years. They could be threaded and worn for decoration, as trophies, or as status symbols. A necklace of large claws from fierce animals showed the prowess of the hunter. In recent times animal claws became popular, mounted in precious metal and made into brooches (Fig. 12.11). The claws usually came from lions or tigers. The practice stopped with the cessation of big game hunting.

Other forms of keratin are bristles, quills and whiskers. The bristles from a sealion or the quills of a porcupine were the type of adornment popular for use on hunting or ceremonial masks. These masks have been used for several hundred years, in various countries, from arctic areas to jungles. Hunting masks were made to look frightening, and were worn in the hope that the prey being hunted would be too terrified to fight back. Bird beaks were another suitable addition to the masks (Fig. 8.16).

Pangolin skin is covered with overlapping scales of keratin, coloured from yellow to brown. They are attached to the skin in a way that allows great freedom of movement. By sewing the scales onto a fabric tunic in a fashion that copied the way they are attached to the animal, they could act as a type of body armour. This was used by the Dyaks in Borneo in the 1800s.

The eggs of exotic birds are still popular as collectors' items, although many are taken illegally from the nests of protected species. However, ostrich eggs are legally traded. These are sometimes cut into discs and pierced through the centre, to be threaded as necklaces (Fig. 12.10) – a form of jewellery popular with some African tribes – but can also be left whole as unusual ornaments. They are cream coloured with a typical, slightly pitted surface. When the eggs were first discovered by explorers, almost 400 years ago, they were brought back to Europe, mounted on ivory, silver or silver gilt stands, and displayed in much the same way as coconuts and calabashes (Fig. 12.12).

13 Plastics

Today plastics are part of everyday life. They are used for such diverse items as children's toys, food packaging, and building materials for houses, aeroplanes, ships and cars. They are very varied, both in their composition, and in their use. The total world sales of plastics is greater than that of metal. Few totally new plastics are nowadays developed, but existing plastics are becoming much more sophisticated, and new uses are found for them. The world of plastics is an ever-changing one.

There are many definitions of the word 'plastic'. It is possibly more correct to use the term 'polymer' when attempting to describe the nature and properties of plastic materials. A polymer is a material that can be softened and moulded. As a material, a polymer is made up of large molecules containing carbon, which are linked together in a chain. The large chains of molecules are made up of short chains of molecules that hook together. This process of connection is called 'polymerisation', and the resulting material is a 'polymer'. On heating, the chains of molecules in some polymers link together – which is called cross-linking – so that when the material cools down it has lost its 'plasticity' – that is, its ability to be moulded or reshaped. Such polymers are called 'thermosets'. Other types of polymer do not cross-link when heated, and the chains of molecules do not attach themselves to each other. These can be reheated and reshaped, each time keeping their new shape when they cool down. They are called 'thermoplastics'. Thermoplasticity can be observed when, for example, some plastic household items are subjected to heat, for example in a dishwasher or oven, and change their shape.

Horn and tortoiseshell can be described as natural polymers, as they can be heated and moulded. Further, as this process can be repeated, they can be called thermoplastics.

Plastic materials are sometimes termed 'resins'. These can be synthetic or natural. Copals are natural resins, some of which can turn into amber, given enough time. This happens by an evaporation of some of the chemicals, while other chemicals cross-link and polymerise, turning a material that was once soft and malleable into one that is hard and stable. In this case the process happens over a period of time, rather than by heating. Some man-made resins also lose their plasticity with time, as they too cross-link, with the result that they can no longer be moulded.

Production methods

As already stated, there are many different methods of processing plastics, according to the material used and the desired finished product. Plastics are wonderful materials for shaping. They can be made into flat sheets, or they can be reinforced with fibres. They can be blow moulded and made into hollow objects such as drinks bottles, or they can be thermoformed for making food cartons. There is a variety of processes for the different synthetic plastics, but all of them start with the required chemicals in pellet, powder or liquid form. These are melted, mixed with additives, heated and shaped.

The three processes that are used for the organics that can be plasticised, or their simulants, are extrusion, compression moulding and injection moulding.

- **Extrusion** is used to produce tubes, continuous sheets, and rods. These can be simple, or have a profile when cut in cross-section. The heated, plasticised material is forced through a shaped nozzle or die in a continuous, long piece. It can later be cut, turned and polished as required.

  This method was used to make 'pressed amber', using inferior amber or scrap that was reconstituted, before being heated and thereby plasticised. An example is 'Ambroid'. Today it is used to produce reconstituted amber beads. The 'amber' is produced in long rods, which are then sliced into short pieces and cut, filed and tumbled to make them rounded, before being polished.
- **Compression moulding** is the oldest method of moulding a plastic material, and has been used for several centuries. The raw material is placed in the bottom half of a mould and forced into shape when the top half of the mould is clamped down upon it. Either the mould or the material is heated, and sometimes both. The process uses large, heavy moulds and is more suitable for large items. It is also more suitable for thermosets, as thermoplastics must cool in the mould to cure, and this increases the cost of production.

  The method was used to shape bois durci, an early, natural polymer, which was made into plaques and large items such as ink wells. Compression moulding was also sometimes used for large Bakelite items.

Horn and tortoiseshell were also moulded by a form of compression moulding. The raw material was heated, and compressed between two halves of a mould, which were also possibly heated. However, as this system involves small moulds and thermoplastic material, some people prefer to refer to it as **moulding by heat and pressure**. The method is still used today for working horn.

- **Injection moulding** is a method used for making solid objects in large numbers. It is used to make many every day items such as buckets, spoons and bowls. The method involves a plasticised material being injected into a mould through a small hole. When the material is sufficiently hard, the mould is taken apart and the resulting object removed. The mould surface can be smooth or patterned.

  Injection moulding is used in the production of inexpensive jewellery in such forms as beads or bangles with patterned surfaces, and items such as brooches and cameos that imitate coral or shell. Other examples of products made this way are the 'ivory carvings' which are often slightly flexible (Fig. 3.11).

PLASTIC TYPES

Modern plastics are synthetic polymers, derived from oil. Early plastics were based on a variety of natural materials, including sugar cane, cotton, rubber and coal tar. The following attempts to give a brief outline of the early plastics most used in connection with organic gem materials.

- **Rubber** is a natural substance that comes from trees of the genus *Hevea*, which grow in tropical regions. Incisions are made in the bark of the tree, and the rubber, in the form of latex, drips into small cups that have been tied to the trunk. It is listed here because it is a mouldable, natural polymer.

  Rubber is soft at room temperature, and flexible. There are indications that it was first used hundreds of years ago by people living locally to its habitat, but by around 1820 it was also being used in Europe for waterproofing clothing. It has, over the years, been used somewhat experimentally in some jewellery and as a decorative material, but it can deteriorate with time and become brittle.
- **Vulcanite** was probably developed by several people at about the same time. It was patented by Hancock in England in 1843, and

by Goodyear in America in 1844. Various people had discovered that, when heated together with sulphur, rubber cross-linked to give a material that retains its flexibility over a wide temperature range. Examples of this are elastic bands and gum boots. The process is called vulcanisation. It was further discovered that, if more sulphur was added, the rubber became very hard. This material was called 'vulcanite' or 'ebonite'. The latter is the term used in the chemical industries, while in the antiques and jewellery world 'vulcanite' is the more usual name. (It is commonly called 'hard rubber' in the United States.)

Vulcanised rubber has been used for many purposes, including household goods, electrical insulation, and car tyres. It is naturally coloured black, so could not be dyed. Being black, it was ideal as a jet simulant. It loses its colour with age, and fades to a greenish-brown hue, caused by its sulphur content (Fig. 2.3). Vulcanite is still in production today.

- **Gutta percha** is another type of natural rubber, though nowadays less well known. It comes from trees of the genera *Palaquium* and *Payena*. It is related to rubber. It is hard at room temperature, but softens at a much lower temperature than vulcanite. Gutta percha deteriorates with age when exposed to air and light. However, it proved ideal for use as insulation for underwater telegraph cables, and was therefore instrumental in the development of this area of communication. It was also used for golf balls and for packing teeth during root canal work.

  There has been some discussion as to whether many of the items simulating jet, and purporting to be vulcanite, are, in fact, made from gutta percha. Experiments were undoubtedly made with the material when it first came into use (among others being vulcanisation), but it was unlikely that very much came of it, as gutta percha was a very good material in its own right. Any attempt at altering an excellent and expensive material, in order to imitate other materials already available – such as vulcanite – would have been unnecessary and superfluous.

  It is possible that the muddle about gutta percha and vulcanite, for use in jewellery, came about because the name 'gutta percha' became a generic term in America, covering many similar materials, including vulcanite and shellac.
- **Shellac** was in use by 1856. It is another natural polymer and is based on a secretion from an insect, *Laccifer lacca* – the lac insect – which is a plant parasite. To produce shellac, the secretion and the insects are scraped off plants and filtered, resulting in a hard, brittle thermoplastic material. It is mixed with wood flour or a mineral filler, pressed, and steamed, dyed, and rolled into sheets to be moulded.

Figure 13.1 Shellac Union case.

Shellac is dark in colour, and is usually dyed black, red or brown. Its best known uses were as gramophone records. Another use was for American 'Union cases', which were small boxes used as folding picture frames (Fig. 13.1). They could be closed to preserve the silver nitrate coated glass photo slides called Daguerreotypes, which deteriorate in the light. Shellac was also used for dressing table sets and for brooches. It is a brittle material and can crack or chip. Shellac is compression moulded, and is still produced today.

- **Bois durci** was also a natural polymer. It was made of albumen, usually from ox blood, but sometimes from egg white, reinforced with wood flour. The wood flour was usually from a dark hardwood, which gave the best result. It was compression moulded in heavy, metal moulds. It is a heavier material than most of the plastics, even among the early ones. It is no longer in production.

  It was developed by François Charles Lepage in France in 1855, and used to make decorative objects, such as picture or mirror frames, and for desk items. It was particularly suitable for making wall plaques, which usually depicted mythological scenes or portraits of famous people (Fig. 13.2). One popular use for the plaques was to mount them on the front of upright pianos.

Figure 13.2 Bois durci plaque of Napoleon.

Figure 13.3 Degraded celluloid box that once imitated tortoiseshell.

- **Cellulose nitrate** is a semi-synthetic plastic based on cellulose from wood or cotton. It is mixed with nitric and sulphuric acids, and uses camphor as a plasticiser. It is another compound that was being developed by various people in different places at the same time, but was launched in England in 1862 as 'Parkesine'. It was later called 'Xylonite'. Cellulose nitrate was finally patented in America in 1870 under the name 'celluloid', but has been known by over 60 different trade names during the years it has been in production.

  It is a highly inflammable material, but was nonetheless used to make stiff, white collars and cuffs for men, and toys, as well as other items such as billiard balls. It proved to be the breakthrough needed in cinematography as it was the base for flexible ciné film.

  Cellulose nitrate was the first of the man-made materials that could be produced in pale colours, and that could be either opaque or transparent. It could be dyed any colour, and given a pearly lustre. Thus it was an ideal material to use when imitating organics, and was used a lot to make jewellery and ornamental items. Amber, coral, shell, tortoiseshell and ivory have all been imitated in celluloid (Fig. 11.22). In the case of ivory, a grain imitating elephant ivory was attained by using opaque and slightly translucent material in alternating layers.

  Unless stored in perfect conditions, an object made of celluloid can degrade, giving off a smell of camphor (which smells like moth balls), and turning sticky or powdery (Fig. 13.3).
- **Cellulose acetate** was developed in 1894 as a safe alternative to the combustible cellulose nitrate. It was made with acetic acid instead of nitric acid. It was not put into commercial production until the 1920s – after the First World War – when it was called 'rayon'. If not kept in ideal conditions it can, in rare cases, degrade, giving off a smell of vinegar.
- **Casein** was a major plastic in the imitation of organic gem materials. It was developed in the late nineteenth century, and was also known by various trade names such as 'Erinoid' and 'Galalith'. It derived from proteins in the curds of skimmed milk, which were moulded and cured with formaldehyde.

  Casein was versatile in that it could take any colour, or could be moulded in a combination of colours, which made it a very attractive plastic. Its most common uses were as knitting needles and fountain pen barrels. The latter often displayed a pearly lustre. Casein is still in use today for button manufacture. It was also made into brightly coloured jewellery, such as linked bangles. It made a good imitation for ivory as it has a very similar fluorescence under ultraviolet light – the only plastic to display this.

  The material is slightly porous and is therefore affected by its surroundings. This can lead to surface crazing but seldom to a complete breakdown.

Figure 13.4 Cast phenolic brooch, imitating amber (magnified).

- **Bakelite** (or **phenol formaldehyde**) heralded the new wave of plastics that were totally synthetic. It was made from phenol and formaldehyde, and was patented in 1907 by Leo Baekeland, under the name of 'Bakelite'. As with other plastics, it took a while to perfect, but finally Baekeland found ways of stabilising the compound so that it could be put into production.

  Bakelite darkens with age, so a pigment or dye was often added in manufacture to mask this effect. It also had a filler added to reinforce the material, giving it an opaque finish. It could be moulded, and has been used for countless numbers of articles, such as radio cabinets, or the well-known, black telephone.

  The decorative form of Bakelite was **cast phenolic**, which used a syrup for moulding. The manufacturing process was slower, but allowed for light, bright and even transparent colours. Reinforcing fillers were not added to cast phenolic.

  Both variations are commonly known as Bakelite, which has become a somewhat generic term for all similar materials.

  Bakelite (as cast phenolic) was used to make brightly coloured jewellery, for example bracelets and brooches. It was also used as a simulant for organics such as jet. It was most successful as a simulant for amber, and it can be difficult to distinguish the two

Figure 13.5 Bakelite ear-ring, imitating burmite.

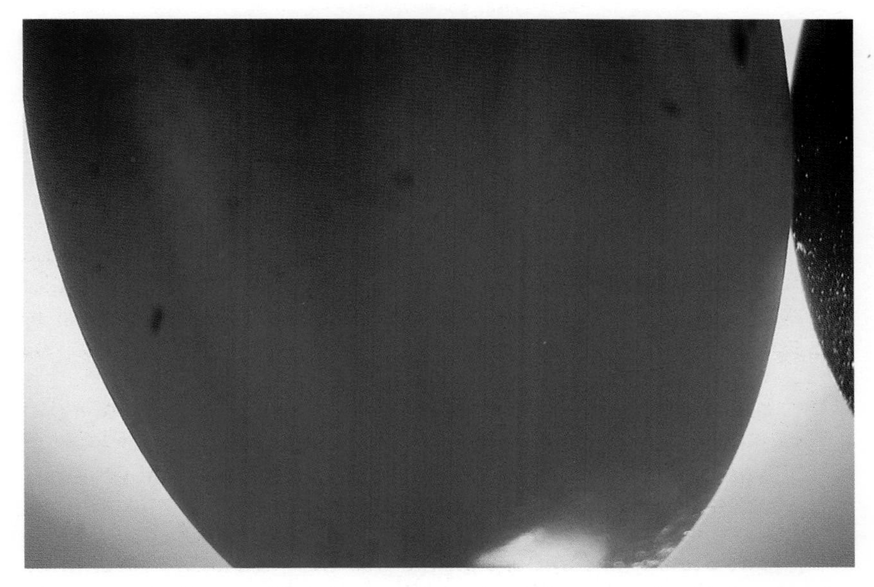

Figure 13.6 Detail of Bakelite ear-ring, viewed by transmitted light, showing faint elongated colour bands.

materials. Pressed amber and Bakelite are especially difficult to differentiate by sight (Fig. 13.4).

When dyed red it has been a convincing simulant of Burmese amber – burmite – a fact that has been much utilised by unscrupulous craftsmen. It is sometimes found for sale in some form of

Figure 13.7 Faked 'beetle in amber' keyring.

jewellery, called 'cherry amber'. There are plentiful examples of 'Chinese burmite carvings', and burmite jewellery which, on investigation, prove to be Bakelite (Figs 13.5 and 13.6).

Bakelite keeps well and does not degrade in the same way as other early plastics, though it can fade. Today it is still in production, and though its uses are limited, it is again gaining popularity. Early Bakelite items are collectors' items, and – to illustrate the change in values – there have recently been reports of dyed pressed amber being sold as Bakelite.

There are a few **modern plastics** that should possibly be mentioned, as plastics are still being used to imitate organic gem materials. It is increasingly difficult, however, to tell exactly which synthetic polymer has been used, and today it is of less importance to know as they are safe and less likely to be inflammable. Also, so far, they are of less value than some of the early plastic imitations, many of which are regarded as collectors' items.

- **Epoxy** is a synthetic polymer which is used in a variety of ways. Strengthened with some form of filler it is used for such purposes as aircraft manufacture. In connection with organics, it is used for items such as imitation scrimshaw, as it can be shaped in pliable moulds which can be peeled off after the polymer has set, minimising the marks left by the mould (Fig. 3.10). It is also used for coatings, sealants and adhesives. It is thermosetting.
- **Polyethylene** is better known as 'polythene'. It replaced gutta percha for covering underwater telegraph cables, and today is produced in greater volume than any other plastic. It is used for countless everyday items such as storage boxes and plastic bags.
- **Polyester** can be produced in a wide variety of forms, from 'Terylene' fibres to car bodies. It is cheap enough to facilitate manufacture of small quantities. Today it is used to make good quality, hand-made plastic jewellery, as it can be produced in opaque or transparent variations, or with a pearly lustre, or filled with metal wire (Fig. 13.7).
- **Polystyrene** is economical in production, and is used to produce low cost items in bulk. It is used to make a lot of inexpensive jewellery such as beads and bangles, and items such as imitation tortoiseshell hair ornaments or combs (Fig. 8.12). It has been used in the production of polybern.
- **Acrylic** has also been used in the production of polybern. Sometimes called 'Perspex', it is a bright, glossy polymer, which has been used for glazing. It has a relatively low chemical resistance.

CARE OF PLASTICS

Far from being indestructible, plastics can, and do, degrade. They are capable of literally falling to pieces (Fig. 13.3). This is true of early and modern plastics, and once the process of degradation has started, it is impossible to reverse.

The most susceptible of the plastics used to simulate organics is cellulose nitrate. Signs of instability and imminent degradation are patches of discolouration, surface crazing or blistering, weeping, and fumes.

There is continuous research into the best methods of preserving plastics, but a few simple guidelines can be given, as follows: in order to try to avoid plastics degrading, they should not be stored in airtight containers, but preferably be loosely wrapped in acid-free tissue paper, and stored in a ventilated environment; they should be kept away from chemicals and solvents, and also from heat and direct sunlight; in the event that they begin to degrade, they should be removed from the vicinity of other items, especially plastics, which they might affect.

14 Quick reference charts

Chart 1: Amber–Copal–Plastics

Materials	*Colour zoning*	*Inclusions*	*Sectility*	*In salt water*	*UV light*	*Burning aroma*
Amber	Swirls	Flora and fauna, sun spangles	Chips	Floats	Variable*	Characteristic*
Copal	Little	Flora and fauna	Chips	Floats	Chalky-blue	Resinous (burns fast)
Pressed reconstituted amber – 'ambroid'	Elongated, feathered swirls	None	Chips	Floats	Faint chalky-blue	Faintly resinous
Pressed chip amber	Blurred blocks	None	Chips	Floats	Faint chalky-blue	Faintly resinous
'Polybern'	Clear blocks	None	Pares	Sinks	Inert	Acrid
Plastics	Swirls	Only in 'fakes'	Pares	Usually sinks	Inert	Acrid

*See Chapter 1, 'Amber and copal'.

Chart 2: Jet and simulants

Material	*Weight*	*Colour*	*Lustre*	*Feel*	*Fracture*	*10× lens*	*Work method*	**Links/ fittings*	*Streak*	*Burning aroma*
Jet (geologically older)	Light	Deep black	High	Warm	Conchoidal	Little, possibly wood grain	Carved	Cut/glued	Brown	Coal
Jet (geologically younger)	Light	Deep black	Varies	Warm	Varies	Little, possibly wood grain	Carved	Cut/glued	Varies	Coal
Other coals	Light	Varies	Mostly high	Warm	Varies	Little, possibly spores	Carved	Cut/glued	Black	Coal
Bog oak	Light	Very dark grey	Dull	Rough, warm	Rough	Wood grain	Carved	Cut/glued	Black	Wood
Horn	Light	Dyed black	Slightly dull	Warm	Rough, flaky	Striations	Heat pressed	Bent/welded	Minimal	Keratin
Vulcanite	Light	Can fade to khaki	Dulls with age	Plasticy, warm	Smooth	Colour spots, surface dimples	Moulded	Bent/screwed	Brown	Rubber and sulphur
Bakelite	Light	Black	High	Plasticy, warm	Smooth	Little	Moulded	Moulded or screwed	Black	Acrid
Celluloid	Light	Black	High	Plasticy, warm	Smooth	Little	Moulded	Moulded or screwed	Black, minimal	DANGER! Combustible
Glass	Heavy	Black, transparent	Vitreous	Cold	Conchoidal	Little	Moulded	Wired	None	None

*'Links' refers to chain and similar links. 'Fittings' refers to metal pins or clasps.

Chart 3: Ivories

Species	*Tooth type*	*Overall shape and size*	*Shape X-section*	*Longitudinal section*	*Cementum X-section*	*Dentine X-section*
Elephant tusk	Upper incisor	Long, curved, max. 3 m	Round or oval	Irregular wavy lines	Thin layer	Intersecting arcs, ('engine turning' pattern)
Elephant tooth	Molar	Large block, ridged top, 30 cm	Rectangular	Folds of dentine, enamel and cementum	Folds of dentine, enamel and cementum	Folds of dentine, enamel and cementum
Mammoth tusk	Upper incisor	Long, more curved than elephant, max. 4 m	Round or oval	Irregular wavy lines	Can be thick, layered	Intersecting arcs, ('engine turning' pattern)
Walrus tusk	Upper canine	Slightly curved, 60 cm	Uneven oval	Can show both forms of dentine	Thin layer	Primary: fine concentric lines. Secondary: bubbly appearance
Walrus tooth	All others	Round, irregular pegs, 5 cm	Uneven round or oval	Can show both forms of dentine	Thick with concentric lines and clear junction with dentine	As tusks, but little Secondary
Sperm whale	All in lower jaw	Slightly curved fat pegs, 20 cm	Oval to round	Lines in dentine, distinct junction with cementum, dark central line	Clearly defined junction with dentine	Marked concentric lines, dark central dot
Orca ('killer whale')	All	Slightly curved, narrower pegs, 15 cm	Round	As for sperm whale	As for sperm whale	As for sperm whale
Narwhal	Upper incisor	Straight, spirals anti-clockwise, 2 m	Round with peripheral indentations	Slight diagonal lines from spiralling, hollow	Clearly defined junction with dentine	Concentric lines
Hippopotamus	Upper canines	Short curved pegs, 40 cm	Oval or round with indentations	Very faint lines	Thin layer	Faint concentric lines, sometimes arched cavity at centre
	Lower canines	Longer, more curved, 70 cm	Triangular	Very faint lines	Thin layer	Faint concentric lines
	Incisors	Peg shaped, 20 cm	Oval to round	Very faint lines	Thin layer	Faint concentric lines, dot at centre
Suids (warthog, etc.)	Upper and lower canines	Curved, longi-tudinal furrows, 10–40 cm	Squared or triangular	Mostly hollow	Thin layer	Irregular faint concentric bands

*The size of tusks and teeth depend upon the age and sex of the animal. Those given are approximate and for purposes of comparison only.

Chart 4: Ivory–Bone–Antler–Plastic

Material	*Colour*	*Transparency*	*Feel*	*Structure: X-section*	*Structure: L-section*	*Flexibility*	*UV light*
Ivory	Cream	Opaque	Warm	May have pattern[1]	May have pattern[1]	Rigid	Chalky-blue
Bone	Cream	Opaque	Warm	Black spots	Black lines	Rigid	Chalky-blue
Antler	Pale beige to grey	Opaque	Warm	Central core	May show core[2]	Rigid	Chalky-blue
Plastics	Any colour	Varies	Warm	May show regular blocks	May show regular stripes	Can be slightly flexible	Mostly inert[3]

1. See Chapter 3, 'Ivory'.
2. See Chapter 5, 'Antler'.
3. Inert, except casein.

Chart 5: Horn–Rhino horn–Tortoiseshell–Plastic

Material	*Colour and pattern*	*Translucency*	*Structure*	*X-polars*	*UV light*
Horn (cattle horn, etc.)	Cream, browns to black in patches	Almost opaque, except if in thin section, bleached or pressed	Longitudinal striations	Colours in fine stripes if bleached or very pale	Slight fluorescence on pale areas
Rhino horn	Mid-brown with fine stripes	Only in thin sections, e.g. of fine carving	Longitudinal striations	Not possible	Mostly inert. Some fluorescence on outer surface
Tortoiseshell	Dappled brown and honey	In light areas only	Microscopic blobs	Coloured spots in pale areas	Slight beige fluorescence in pale areas
Blond tortoiseshell	Even honey	Translucent	Minimal	Coloured spots	Slight beige fluorescence
Plastic	Varies	Varies	None	Possible swirls of colour	Inert except casein

Chart 6: Pearls and simulants

Material	*Shape*	*Surface*	*Drill hole observations*
Natural pearl	Can be irregular, seldom well matched	Slightly gritty but looks smooth Microscopic over-lapping platelets	No sign of bead nucleus
Nucleated cultured pearl	Almost regular, well matched. Also special shapes	As for natural pearl	Bead nucleus, possible signs of dye accumulation or different coloured bead
Shell bead simulant	Very regular	Very smooth, no sign of microscopic overlapping platelets. Possible signs of parallel layers	Possible signs of dye
Glass filled simulant	Very regular	Glassy, impression of double surface	Signs of wax or resin filling
Plastic coated bead simulant	Very regular	Not gritty, but looks slightly rough. Possible surface peeling	Signs of surface chipping

Note: non-nucleated cultured pearl, see Chapter 9, 'Pearl'.

Chart 7: Coral–Shell–Pearl–Plastic–Glass

Material	*Feel*	*Structure**	*Surface pattern*	*Flexibility*	*UV light*
Precious coral	Cold	Concentric rings in cobweb pattern, and longitudinal striations	See structure	Rigid	Pale colours: chalky-white
Shell	Cold	Striations, possibly different colour layers (plus microscopic striations)[2]	Flame pattern, striations, or microscopic platelets (mother-of-pearl)	Rigid	Pale colours: chalky-white[2]
Pearl[1]	Cold	As surface pattern	Platelets follow surface shape, possible surface blemishes	Rigid	Chalky-white to yellowish
Coated plastic[1]	Warm	None	None	Varies	Varies
Glass	Cold	None	None	Rigid	Varies

*As possibly seen at the end of a bead.
1. Not including glass or plastic pearl simulants.
2. See Chapter 10, 'Shell'.

Glossary

10× lens A small hand lens which magnifies at ten-power. Much used in gemmology.

Aragonite A crystal form of calcium carbonate.

Burning test A test involving burning a minute scraping from a specimen to judge its melting point and smell.

Cabochon A domed gemstone shape.

Calcite A crystal form of calcium carbonate.

Cameo A type of relief carving on a flat background, used in jewellery.

CITES The Convention on International Trade in Endangered Species of Wild Fauna and Flora. An international agreement between governments.

Composite An item which is composed of parts, usually one natural and one or two of a synthetic material. The natural material may be capped, backed or filled to add bulk or protection.

Conchoidal fracture The ridged or shell-shaped pattern which occurs when some materials break, e.g. amber or jet.

Cretaceous A geological period, 144 to 65 million years ago.

Crossed-polars See Polariscope.

Discoidal stress cracks Disc-shaped cracks caused by stress in a material such as amber, when it is heated and cooled. Sometimes called 'sun spangles' or 'lily pads'.

Doublet An item made of two parts. See Composite.

Drill hole The hole drilled through the middle of a pearl or bead to enable stringing.

Eglomisé A technique involving the application of etched foil to the back of transparent amber as decoration.

Family See Taxonomic table.

Filler A substance added to a powdered or liquid material, e.g. plastic, to give extra bulk or strength.

Fluorescence Normally refers to the visible light given out when a specimen is held in X-ray, UV or some visible light (e.g. sunlight).

It is caused by the specimen absorbing energy in one wavelength and re-emitting it in another.

Genus see Taxonomic table.

Gorgonin A horn-like hard proteinaceous material, similar to keratin, that makes up all or part of some coral skeletons.

Hot-point test A test involving pressing the tip of a hot needle against a specimen to judge its melting point and smell.

Imitation A material imitating another one. A simulant.

Inclusion An item or cavity enclosed in a specimen, e.g. an insect inclusion in amber.

Inert Does not react. The term may refer to chemical or physical reactions, actions or processes.

Infrared spectroscopy The measurement of the absorption of light energy, in infrared light, of a tiny scraping taken from a specimen, to determine its nature or origin.

Inorganic A material not originating from plants or animals.

Jurassic A geological period: 213 to 144 million years ago.

Keratin A protein containing large amounts of sulphur-rich amino acids. It is a material vital to the animal kingdom as it makes up horn, hoof, baleen, claws, hair, fur and human nails, and is also present in skin.

Light diffraction The splitting of white light into its component parts, i.e. the spectrum, when it passes through a small aperture.

Light interference The extinction of light, or the predominance of one or more colours, caused by rays of light which are travelling in the same path but are out of phase, interfering with each other.

Lustre The amount and quality of light reflected from a specimen.

Melanin A dark pigment found in hair, skin, etc.

Miocene A geological time, 25 to 5 million years ago.

Neolithic The last period of the Stone Age.

Netsuke A small Japanese carving, worn on the clothes.

Order See Taxonomic table.

Organic In gemmology, a material derived from or produced by living animals or plants. In chemistry 'organic' denotes a compound of carbon.

Osseous Made of bone.

Pare To slice very thinly. To peel.

Paleolithic The earliest period of the Stone Age.

Phylum See Taxonomic table.

Piqué A technique involving the inlaying of metal in tortoiseshell or horn.

Play-of-colour Iridescence due to light interference or diffraction.

Polariscope An instrument used to test the reaction of a specimen to light passed through two polarising filters, adjusted to block out all light, with the specimen placed between them.

Radiolucent Allows X-rays to pass through.

Radio-opaque Does not allow X-rays to pass through.

Raw Material in its unworked form.

Reflected light Light which is reflected at the surface of a specimen, or from an internal fracture (e.g. 'sun spangles' in amber).

Refractive index (RI) A measure of the degree by which a material refracts (bends) visible light entering it from the surrounding medium (air). A much used gemmological test.

Rough Material in its unworked form.

Sectility The capability of being cut with a sharp knife, without breaking. See Pare.

Simulant A material which simulates a natural one. An imitation.

Specific gravity (SG) The ratio of the mass of a substance to the mass of an equal volume of water at a temperature of 4°C. A test much used in gemmology.

Streak The colour of a material in its powdered state, which may differ from that of the mass of the material.

Streak test A test to obtain a sample of powder from a specimen in order to judge its colour. It can be carried out by rubbing the specimen on a hard rough surface of neutral colour.

Striations Fine lines which are parallel, or almost parallel.

Taxonomic table A table classifying living organisms. It starts with 'Kingdom', e.g. 'animal', and subdivides through Phylum, Class, Subclass, Order, Family, Genus and Species. The genus and species are the most used and are always written in italics. (Species is occasionally termed 'Name'.)

Tertiary A geological period, 65 million years ago to the present day.

Thermoplastic A material that can be heated and moulded repeatedly, and retains its shape each time it is cooled.

Thermosetting A material that can be heated and moulded and retains its shape after cooling, but which cannot be reshaped or moulded again on reheating.

Transmitted light Light which is passed through a specimen.

Triplet An item made of three parts. See Composite.

Ultraviolet light Light in wavelengths beyond violet light, which is invisible to the human eye, and which causes some materials to fluoresce. Gemmological tests usually use both short wave UV and long wave UV, but as organics react much more strongly to long wave, and many are inert to short wave, this book refers to long wave only.

UV light See Ultraviolet light.

Vitreous lustre High glassy lustre.

X-ray Electromagnetic radiation of very short wavelength, which can pass through some substances which are opaque to visible light.

Bibliography

Alderton, David (1993). *Turtles and Tortoises of the World.* Blandford Press.

Allason-Jones, Lindsay (1996). *Roman Jet in the Yorkshire Museum.* The Yorkshire Museum.

Bradley Martin, Esmond and Chrysee (1982). *Run, Rhino Run.* Chatto and Windus.

Brown, Grahame and Moule, A.J. (1982). Structural Characteristics of Various Ivories. *Gemmological Association of Hong Kong.* Vol. V.

Brown, Grahame (1986). Gemmology of the Cameo Shell. *Australian Gemmologist.*

Brown, Grahame (1988). Paua Shell. *Australian Gemmologist.*

Burak, Benjamin (1984). *Ivory and Its Uses.* Charles E. Tuttle Co. Inc.

Doubilet, David (1996). *Pearls. From Myths to Modern Pearl Culture.* Schoeffel Pearl Culture.

Edwards, Hugh (1994). *Pearls of Broome and Northern Australia.* Private publication.

Eltringham, S.K. (1999). *The Hippos.* Poyser Natural History, Academic Press.

Espinoza, Edgard O. and Mann, Mary-Jacque (1992). *Identification Guide for Ivory and Ivory Substitutes.* WWF.

Ezra, Kate (1984). *African Ivories.* The Metropolitan Museum of Art.

Faber, Ole, Frandsen, Lene B. and Ploug, Mariann (2000). *Rav.* Museet for Varde By og Omegn.

Fabricius, Katharina and Alderslade, Philip. (2001). *Soft Corals and Sea Fans.* Australian Institute of Marine Science.

Farn, Alexander E. (1990). *Pearls, Natural, Cultured and Imitation.* Butterworth and Co.

Fisher, Angela (1984). *Africa Adorned.* The Harvill Press.

Fosdick, Peggy and Sam (1994). *Last Chance Lost?* Irvin S. Naylor.

Fraquet, Helen (1987). *Amber.* Butterworth and Co.

Grasso, Tony (1996). *Bakelite Jewelry.* The Apple Press.

Greep, S.J. (1987). Use of bone, antler and ivory in the Roman and medieval periods. *Archaeological Bone, Antler and Ivory.* UKIC.

Grimaldi, David A. (1993). The Care and Study of Fossiliferous Amber. *Curator.* **36/1**, 31–49.

Grimaldi, David A.; Shedrinsky, Alexander; Ross, Andrew *et al.* (1994). Forgeries of Fossils in 'Amber': History, Identification and Case Studies. *Curator.* **37/4**, 251–274.

Grimaldi, David D. (1996). *Amber, Window on the Past.* American Museum of Natural History.

Hansen, Keld (1996). *Legetøj i Grønland.* Nationalmuseet.

Hardwick, Paula (1981). *Discovering Horn.* Lutterworth Press.

Hayward, Bruce W. (1989). *Kauri and the Gumdiggers.* The Bush Press.

Hickson, Sydney J. (1924). *An Introduction to the Study of Recent Corals.* Manchester University Press.

Impey, O.R. and Tregear, M. (1983). *Oriental Lacquer.* Ashmolean Museum.

Iverson, John B. (1992). *A Revised Checklist with Distribution Maps of the Turtles of the World.* Private publication.

Jackson, Beverley (2001). *Kingfisher Blue. Treasures of an Ancient Chinese Art.* Ten Speed Press.

Jaffer, Amin (2002). *Luxury Goods from India. The Art of the Cabinet-Maker.* V&A Publications.

Jensen, Jørgen (1982). *Rav, Nordens Guld.* Gyldendahl.

Katz, Sylvia (1994). *Early Plastics.* Shire Publications Ltd.

Kenny, Adele (2001). *Photographic Cases: Victorian Design Sources 1840–1870.* Schiffer Publishing Ltd.

Kunz, George Frederick and Stevenson, Charles Hugh (1908). *The Book of the Pearl. The History, Art, Science and Industry of the Queen of Gems.* The Century Co., New York.

Landman, Neil H.; Mikkelsen, Paula M.; Bieler, Rüdiger *et al.* (2001). *Pearls: A Natural History.* American Museum of Natural History, and The Field Museum.

Lauer, Keith and Robinson, Julie (1999). *Celluloid. Collector's Reference and Value Guide.* Schroeder Publishing Co. Inc.

Lister, Adrian and Bahn, Paul (1995). *Mammoths.* Boxtree Ltd.

Liverino, Basilio (1989). *Red Coral, Jewel of the Sea.* Analisi.

MacGregor, Arthur (1985). *Bone, Antler, Ivory and Horn. The Technology of Skeletal Materials Since the Roman Period.* Croom Helm.

Mack, John; Barry, Iris; Scarce, Jennifer *et al.* (2002). *Ethnic Jewellery.* The British Museum Press.

Maskell, Alfred (1905). *Ivories.* Methuen and Co.

Matthews, L. Harrison (1978). *The Natural History of the Whale.* Weidenfeld and Nicholson.

McManus, Michael (1997). *A Treasuey of American Scrimshaw. A Collection of the Useful and Decorative.* Penguin Studio.

Miller, Anna M. (1991). *Cameos Old and New.* Van Nostrand Reinhold.

Morton, J.E. (1979). *Molluscs.* Hutchinson University Library.

Muller, Helen (1987) *Jet.* Butterworth and Co.

National Research Council (1990). *Decline of the Sea Turtles. Causes and Prevention.* National Academy Press.

Nyborg, Preben (1993). *Design with Plastics.* Danish Design Centre.

O'Connor, S. (1987) The Identification of Osseous and Keratinaceous Materials at York. *Archaeological Bone, Antler and Ivory.* UKIC.

O'Connor, T.P. (1987) On the Structure, Chemistry and Decay of Bone, Antler and Ivory. *Archaeological Bone, Antler and Ivory.* UKIC.

Penniman, T.K. (1952). *Pictures of Ivory and other Animal Teeth, Bone and Antler.* Pitt Rivers Museum, University of Oxford.

Petersen, David (1991). *Racks. The Natural History of Antlers and the Animals that Wear Them.* Capra Press.

Poinar, George O. (1992). *Life in Amber.* Stanford University Press.

Quye, Anita and Williamson, Colin (1999). *Plastics: Collecting and Conserving.* NMS Publishing Ltd.

Rice, Patty C. (1987). *Amber, the Golden Gem of Ages.* The Kosciuszko Foundation Inc.

Ritchie, Carson I.A. (1970). *Carving Shells and Cameos.* Arthur Barker.

Roesdahl, Else (1995). *Hvalrostand, elfenben og nordboerne I Grønland.* Odense Universitetsforlag.

Rosing, Jens (1986). *Havets enhjørning.* Wormianum.

Ross, Andrew (1998). *Amber, the Natural Time Capsule.* The Natural History Museum, London.

Ross, Doran H.; Shoshani, Jeheskel; Van Couvering, John A. *et al.* (1992). *Elephant. The Animal and its Ivory in African Culture.* Fowler Museum of Cultural History, University of California.

Schabilion, Shirl (1989). *All in a Nutshell. The Story of the Vegetable Ivory Nut.* The Mississippi Petrified Forest.

Taburiaux, Jean (translated by Ceriog-Hughes, David) (1985) *Pearls, their Origin, Treatment and Identification.* NAG Press.

Vermwij, Geerat J. (1993). *A Natural History of Shells.* Princeton University Press.

Vickers, Michael; Chesney, Charlotte; Lasko, Peter *et al.* (1987). *Ivory: A History and Collector's Guide.* Thames and Hudson.

West, Janet and Credland, Arthur G. (1995). *Scrimshaw: The Art of the Whaler.* Hutton Press Ltd.

Williamson, George C. (1938). *The Book of Ivory.* Frederick Muller Ltd.

Wood, Elizabeth M. (1983). *Corals of the World.* TFH Publications, Inc.

Index

(Page numbers in **bold type** indicate the location of illustrations.)